40 YEARS OF RESTORATION AT TIJUANA ESTUARY, CALIFORNIA

LESSONS LEARNED

PREPARED FOR:

Coastal Conservancy

California State Coastal Conservancy
1515 Clay Street, 10th Floor, Oakland, CA 94612

PREPARED BY:

NORDBY BIOLOGICAL CONSULTING

Chris Nordby, Nordby Biological Consulting
5173 Waring Road # 171, San Diego, CA 92120

IN ASSOCIATION WITH:

SWIA

Southwest Wetlands Interpretive Association
700 Seacoast Drive # 108, Imperial Beach, CA 91932

Tijuana River National Estuarine Research Reserve

Tijuana River National Estuarine Research Reserve
301 Caspian Way, Imperial Beach, CA 91932

SEPTEMBER 2018

NIGHT STAR *Publisher*

NightStarPublisher.com • San Diego, CA

TABLE OF CONTENTS

SECTION.. PAGE

INTRODUCTION

TIJUANA ESTUARY, located in the southwestern corner of the continental U.S. (Figure 1, next page), has a complex history; ecologically, politically and geographically. Ecologically, the estuary has relatively large natural areas, although these tend to be modified and managed. Politically, the estuary has survived attempts at development and is currently managed by a number of federal, state and local agencies and entities. Geographically, Tijuana Estuary is surrounded by the cities of San Diego, Imperial Beach and Tijuana and has been negatively impacted by land use practices on both sides of the border. Because of this high density human population, inadequate sewer systems, unstable slopes and soils, and agricultural activities, Tijuana Estuary has been plagued with water quality and sedimentation problems. These problems were accelerated in the 1980s associated with the rapid growth of Tijuana, Mexico. Occasional sewage spills from Tijuana's sewerage system became continuous flows resulting in reduced salinities in the estuary that was formerly dominated by seawater and commensurate declines in fishes and invertebrates of the tidal channels and creeks. Brackish marsh replaced existing salt marsh habitat to the detriment of some species that rely on salt marsh. Toxic materials were introduced to the water, sediments and biota, although the effects were difficult and expensive to document. Sediment from large scale development of Tijuana was deposited in the estuary, particularly in the south arm, burying tidal channels and salt marsh and converting them to weedy upland habitats. During this period, the need for corrective measures to preserve and restore Tijuana Estuary was identified. Beginning with small scale restoration efforts and progressing

FIGURE 1. REGIONAL LOCATION MAP

to larger projects, land managers and scientists endeavored to reverse the loss of valuable wetlands.

Tijuana Estuary is unique among southern California lagoons and estuaries in that major highways and railroads were built well inland from the site. Thus, Tijuana Estuary escaped the dissection and fill of tidal channels that these infrastructure projects have imposed on other coastal wetlands in the region. This lack of dissection and filling, coupled with other physical factors, has resulted in a system that has remained tidally flushed for the most part as opposed to other systems where inlet closure is more frequent. Conversely, Tijuana Estuary has been affected by land uses that impact most Southern California coastal wetlands. It has been filled and developed around its periphery. It is influenced by urban land uses and is located downstream of two major metropolises. It receives runoff from agricultural fields. Its natural adjacent upland habitats and transition from wetland to upland have been disturbed and developed.

Its protective dune system has been destabilized. The rivers of its watershed have been dammed for many decades. It has been invaded by exotic plant and animal species. Its historic tidal prism has been diminished by chronic and episodic sedimentation. It has been impacted by chronic and episodic sewage flows. More recently, construction of a new border fence has resulted in habitat loss and fragmentation as well as providing an additional source of sedimentation. With predicted sea level rise, the estuary's ability to function as it does currently will be tested. As demonstrated in later sections, each of these stressors have been identified by resource managers and scientists who work at Tijuana Estuary and measures have been implemented to control and reverse their effects on the ecosystem.

Today, many of the sources of sediment and wastewater are more controlled than they were historically, with varying degrees of efficiency. An international wastewater treatment plant was constructed near the U.S.–Mexico boundary in 1997 and during the same period the City of Tijuana undertook a major upgrade of its sewerage system. However, occasional sewage flows continue, especially during the rainy season when pump stations are overwhelmed. Sedimentation basins were constructed in 2005 to capture the major source of sediment from Tijuana, yet challenges remain for disposal of the captured sediment. Numerous habitat restoration projects were successfully implemented and more are in the planning process, yet the losses from past practices have not been recovered fully.

Still, the estuary remains altered from land use practices. The destabilized dunes have been subjected to overwash during winter storms with sand deposited in channels resulting in loss of habitat and tidal prism. Former, natural non-tidal ponds were converted to sewage treatment ponds and, eventually, restored to tidal wetlands. Land uses in the southern arm of the estuary have altered the landscape through filling of wetlands and compaction of soils that resist establishment of vegetation. Sediment deposition has raised the elevation of the salt marsh and tidal creeks in the south arm resulting in conversion to upland habitats. These impacts are presented in greater detail in subsequent sections of this report.

The recently completed study of the historical ecology of the Tijuana River and estuary (SFEI 2017) provides a context for comparisons with present conditions, including former and current land uses, habitat distributions and configurations, river flow, ground water levels, inlet conditions and other factors. This study is referenced throughout this report.

The objective of this report is to summarize restoration activities at Tijuana Estuary during approximately the last 40 years (1976–2016) with an emphasis on the lessons learned as they may inform future restoration efforts and resource management decisions. The intended audience includes resource managers and regulatory agencies, the Tijuana River Valley Recovery Team, Tijuana River National Estuarine Research Reserve staff, future funding agencies and

FIGURE 2. WATERSHED MAP

SITE DESCRIPTION

the general public. This report will also serve to memorialize a comprehensive list of restoration projects in a single document.

Tijuana Estuary is located in the southwestern-most corner of the continental U.S in San Diego County. While the estuary is located entirely within San Diego County, three fourths of its watershed is within Mexico (Figure 2) and is, thus, beyond the control of U.S. interests. The Tijuana River is approximately 120 miles (193 kilometers) long and crosses the U.S.–Mexico boundary approximately five miles east of the estuary. The river is contained within a concrete-lined channel as it flows through the City of Tijuana. As it crosses the border, the river flows through a parcel of land owned by the International Boundary Water Commission (IBWC) and is kept free of vegetation for flood control purposes. Once past the IBWC parcel, the river valley supports riparian vegetation until reaching the tidal salt marshes of the estuary.

There are four dams on the Tijuana River and its tributaries, two in the U.S. and two in Mexico. This, coupled with diversion of polluted water from the concrete channel in Tijuana, results in little flow in the river except during and immediately after rainfall events, or resulting from failures in the sewage or water lines in Mexico. As a result, the estuary is usually dominated by sea water and supports organisms adapted to saline waters and hypersaline soils. As presented later in this document, disruption and alteration of tidal influence at Tijuana Estuary has resulted in long-term effects on the flora and fauna of the estuary.

The coastal climate of Southern California has been compared to the Mediterranean region of Europe with cool, wet winters and warm, dry summers. The coastal vegetation is similar to that of southern France, southern Africa and southwestern Australia. This similarity is based on the timing and amounts of rainfall and river flows, rather than average rainfall. In Southern California, more than 90% of the mean annual precipitation occurs during a six-month period between November and April. It is a climate of extremes, with years or decades of persistent drought sometimes followed by years with torrential floods. It has often been said that there is no "normal" year, in terms of precipitation, for the region. Rather, cycles of wet and dry years are evident in analyses of annual rainfall in the San Diego area (Zedler and Nordby 1986; Zedler et al. 1992).

Rainfall data for the San Diego area cited by Zedler and Nordby (1986) extend back to 1880 and reveal a pattern of relative drought interrupted by wet years in 1883, 1921, 1940, 1951, 1978, 1980 and 1983. More recently, the 2004–2005 rainy season produced the second highest rainfall total in San Diego history. Rainfall patterns varied greatly during wet years with summer storms in some years and winter drought in others. The factors most important to the estuary are not necessarily rainfall totals but the amount and timing of rainfall and stream flow.

The estuary consists of a northern arm, known as Oneonta Slough, a central estuary dominated by the Tijuana River, and a southern arm that follows one of the historic paths of the Tijuana River (Figure 3). The Tijuana River has changed course several times in recent history. Old river channels now serve as estuarine tidal channels in the central portion of the estuary, although these too have been affected by sedimentation. The tidal inlet is located roughly in the central estuary and is primarily influenced by tidal action during dry periods and light to moderate rain events, and by riverine processes, e.g., scour and sedimentation, during and immediately after flood events.

The northern arm of the estuary has largely escaped the wetland filling activities associated with agriculture and military uses more common in the southern arm of the estuary (Zedler and Nordby 1986; Zedler et al. 1992). The northern arm of the estuary and portions of the central estuary have been designated as the Tijuana Slough

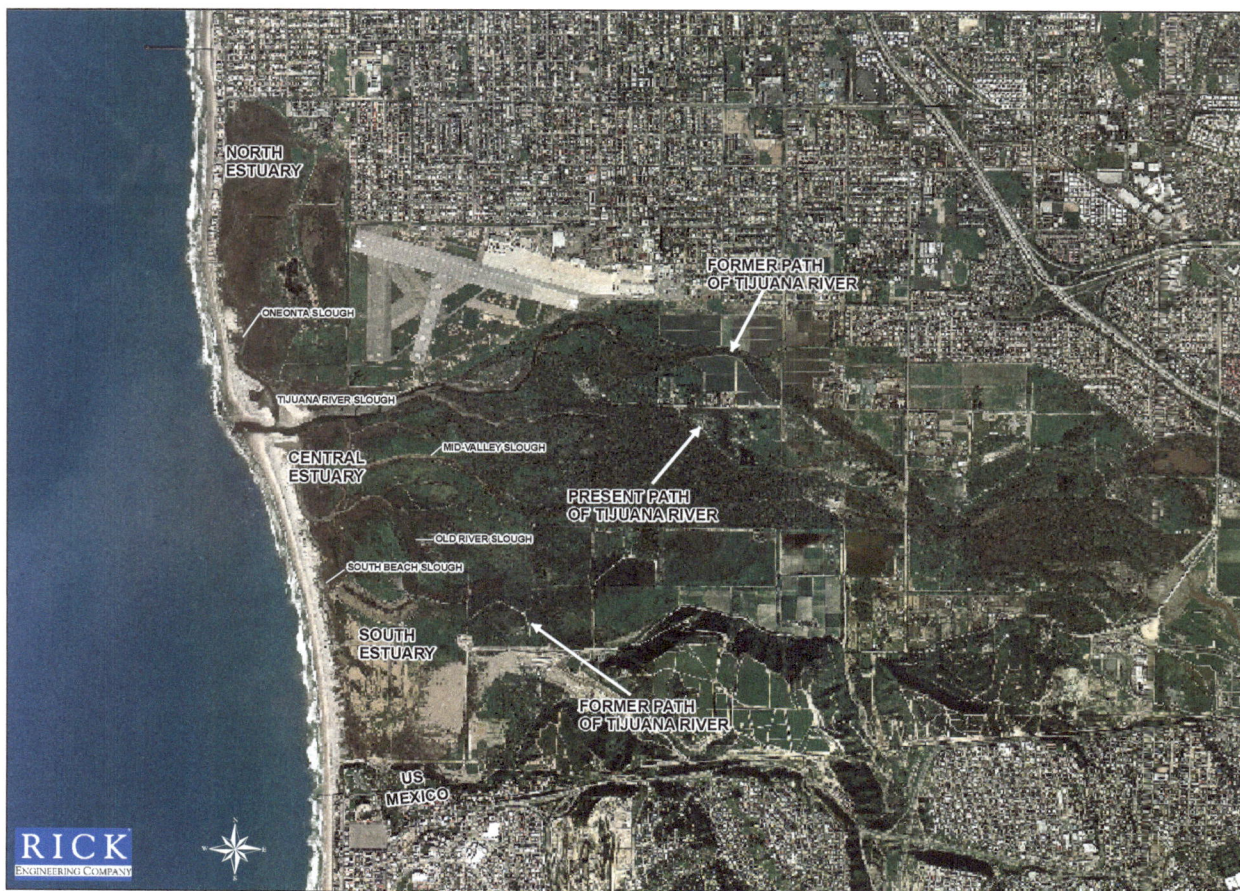

FIGURE 3. AERIAL OVERVIEW OF TIJUANA ESTUARY

National Wildlife Refuge managed by the U.S. Fish and Wildlife Service. The U.S. Navy also owns a large portion of the central and northern arms where it operates the Imperial Beach Naval Outlying Landing Field (Naval Outlying Landing Field; Figure 4).

The southern arm of the estuary has been heavily impacted by filling and diking for agriculture and military use, and more recently, by sedimentation borne across the border from Mexico. The southern border of the estuary, directly north of the Mexican border, consists of a series of coastal mesas and bluffs. These former marine terraces are highly erosive and sediment is easily mobilized by relatively small rainfall events. Much of the south arm was used as a training field by the U.S. Navy during World War II and is currently designated as Border Field State Park (Figure 4), a feature of the California State Parks system.

Tijuana Estuary was designated a National Estuarine Sanctuary in 1982 by the National Oceanic and Atmospheric Administration (NOAA) and later renamed the Tijuana River National Estuarine Research Reserve

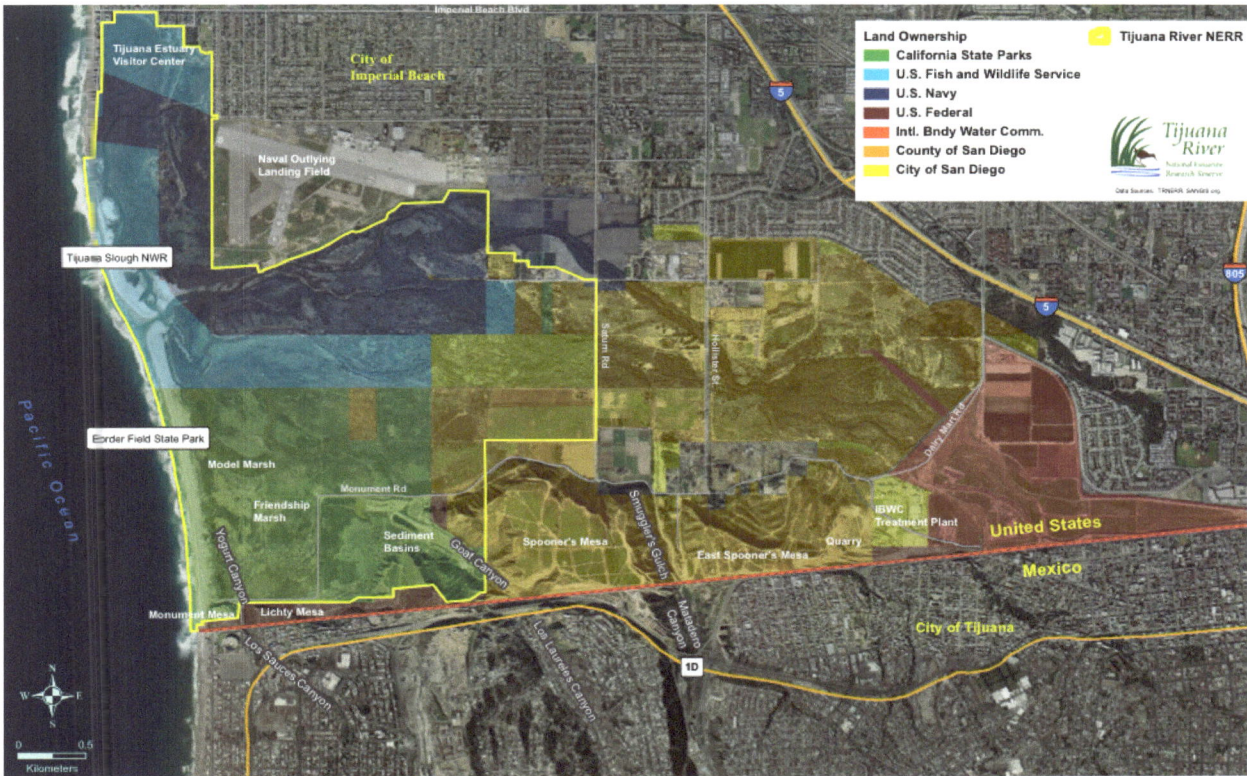

FIGURE 4. TIJUANA RIVER VALLEY OWNERSHIP

(TRNERR). The TRNERR encompasses approximately 2,531 acres of tidal and non-tidal land extending north from the U.S.–Mexico border (Figure 4). The Research Reserve includes the Tijuana Slough National Wildlife Refuge, Border Field State Park and the undeveloped portion of the Naval Outlying Landing Field. The TRNERR was created with a focus on habitat restoration and it has facilitated many of the restoration projects presented in this report.

Numerous institutions have contributed to research and restoration at Tijuana Estuary over the past several decades. These include the National Oceanic and Atmospheric Administration (NOAA); the U.S. Fish and Wildlife Service (USFWS); the U.S. Environmental Protection Agency (EPA); the International Boundary and Water Commission (IBWC); the U.S. Navy; California State Parks (CSP); the California State Coastal Conservancy (Coastal Conservancy); San Diego State University (SDSU); the San Diego Regional Water Quality Control Board (RWQCB); Scripps Institution of Oceanography (SIO), University of California Sea Grant; the County of San Diego and the City of San Diego. These entities are acknowledged accordingly in the sections that follow. Many of the restoration projects were administered by the Southwest Wetlands Interpretive Association (SWIA), a non-profit

organization dedicated to the preservation and restoration of Tijuana Estuary and other regional wetlands.

The objective of this report is to summarize restoration activities at Tijuana Estuary during approximately the last 40 years (1976–2016) with an emphasis on the lessons learned as they may inform future restoration efforts and resource management decisions. The intended audience includes resource managers and regulatory agencies, the Tijuana River Valley Recovery Team, Tijuana River National Estuarine Research Reserve staff, future funding agencies and the general public. This report will also serve to memorialize a comprehensive list of restoration projects in a single document. It is anticipated that the projects described herein will continue to offer new restoration lessons, and that future projects will still further expand these lessons learned. It is envisioned that this report will provide the foundation for a web-based platform that captures key elements of projects completed to date and is updated as new information becomes available.

IMPACTS FROM LAND USE PRACTICES

Because it is important to illustrate the changes that have resulted in the need for large-scale restoration, a brief history of land use and disturbance in the area is presented. Much of the land in the river valley was farmed and later abandoned. A 1953 black-and-white aerial photograph (Figure 5) shows active farming within the central and southern estuary. Cultivated fields are shown very near the coast, suggesting that the ground water was close to the surface and relatively fresh, and that seawater had not intruded far inland. The Tijuana River appears to be active in the center of the river valley as evidenced by sand deposition, as opposed to major channels located on the north and south ends of the valley. More recently, the water table in the project area has fluctuated from within a few feet of the surface with salinities of one–two parts per thousand (ppt) in 1986–1987 at the Pacific Estuarine Research Laboratory (PERL) operated by SDSU (C. Nordby, unpublished data) to more than 15 feet below the ground surface with salinities greater than 38 ppt at Goat Canyon in 2002 (Tierra Environmental Services [Tierra] 2002). Currently, large-scale agricultural uses are confined to the eastern end of the Tijuana River Valley (Figure 4).

Historically, groundwater levels in the lower river valley varied from about one–two meters below the surface in the 1860s–1870s to nine–15 meters below the surface at the peak of groundwater extraction in the 1950s–early 1960s (SFEI 2017). It was estimated that groundwater levels were below the ocean level in the 1950s–1960s period, with sea water intrusion occurring in a large portion of the lower river valley. By the mid-1960s, groundwater extraction had declined and levels were at or near historic highs.

Military activity was also evident within the estuary in 1953 (Figure 5). The U.S. Navy began acquiring Border Field property in the 1930s for use as an auxiliary aviation field. During World War II, the Navy utilized approximately 100 acres as the Naval Operating Base Border Field which was administered from nearby Ream Field (Tierra 1999). During the 1940s, the project area was used as an aircraft gunnery range, machine gun training center, bombing target, air-to-air gunnery field, and emergency landing field. By 1944, 35 buildings existed in the project area including barracks, mess halls, galleys, and support buildings to accommodate 90 enlisted men and six officers. These structures can be seen in the 1953 aerial along a north–south road at the western edge of the agricultural fields. Former military use is also evidenced by a series of roughly circular areas located in the southwest corner of the photograph. In historical accounts of Border Field, in particular the *History of Border Field State Beach* (Donald W. Nicol, State Historian, as cited in Tierra Environmental Services 1999) these areas are alluded to as "rabbit tracks":

"Pilot gunnery practice was given with steam driven targets that dashed about among the dunes on rails called 'rabbit tracks'. Air to air gunnery practice was held using unmanned airplanes until the 1950s when a riddled plane crashed into a barn and killed several race horses."

In 1961, Border Field was deactivated and was turned over to the Navy Electronics Laboratory that conducted experiments there. In 1971, the facility was transferred to the State of California and the buildings were subsequently razed. By 1966, a racetrack for

rehabilitating horses had been constructed and a number of buildings are shown just west of the oval and near what is now the lower parking lot for Monument Mesa (Figure 6). It is assumed that these buildings were associated with the racetrack.

The effects of urbanization are many, including construction of dams, intentional filling of wetland and transitional habitats, diking of tidal channels, sewage and sediment inflows from Mexico, destabilization of the barrier dunes, and loss of tidal prism. These are presented in more detail herein.

IMPACTS FROM FLOODING AND SEDIMENTATION

Flooding of the Tijuana River. Examination of streamflow in the Tijuana River and precipitation records indicates that for approximately 35 years prior to the late 1970s Tijuana Estuary had experienced a relatively stable period in terms of flooding and associated sedimentation. Studies of the estuary during that period showed that fishes and invertebrates typical of marine-dominated systems flourished and recreational fishing and clamming was common. This benign period ended in 1978 and 1980 when El Niño conditions resulted in heavy rainfall and major flooding. In January and February of 1980, floods that exceeded all previous streamflow records simultaneously scoured and filled different parts of the estuary.

Flow gauges on the Tijuana River near the

FIGURE 5. 1953 AERIAL PHOTOGRAPH OF TIJUANA ESTUARY

estuary recorded flows between January and December 1980 that were almost 20 times the average annual volume (Zedler et al. 1992). The Tijuana River shifted its course to the north before flowing west toward the tidal inlet. Major sedimentation occurred in the south arm of the estuary from trans-border tributaries to the Tijuana River. Another major flood occurred in 1983, again associated with El Niño conditions. Rainfall

1966

Pacific
Ocean

Race Track

Structures

Old River
Slough

Monument Road

N

NOT TO SCALE

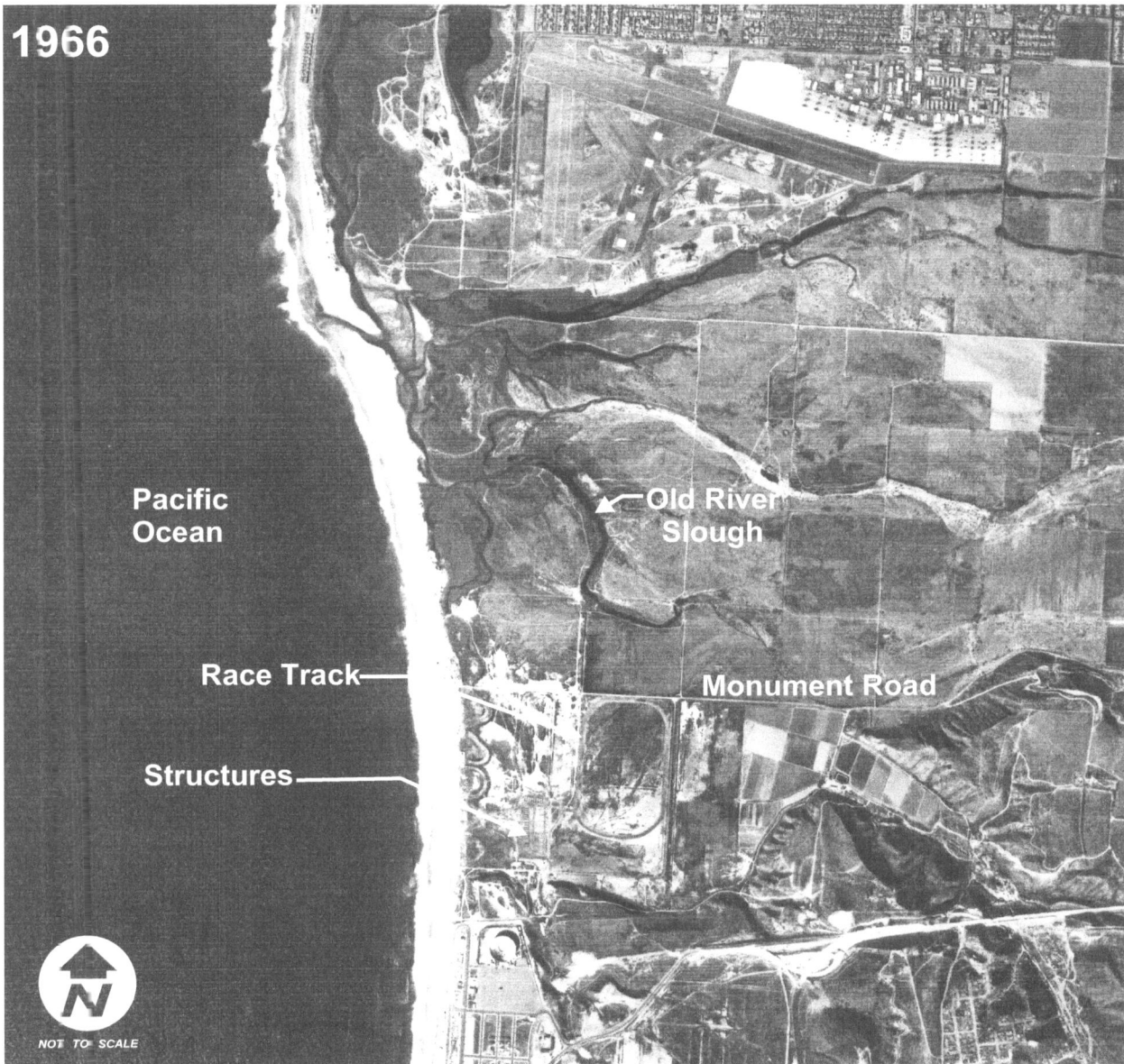

FIGURE 6. 1966 AERIAL PHOTOGRAPH OF TIJUANA ESTUARY

was 2.4 to 3.8 times greater than average for the period of February through April 1983. Rodríguez Dam and Reservoir, situated on the Tijuana River in Mexico approximately 11 miles east of the City of Tijuana, received inflows that were greater than 10 times the average during 1983, requiring release of impounded water. Reservoir release continued through summer and fall of 1983 resulting in high flows to the estuary for the entire year.

The impacts to the estuary from flooding were far reaching. Sediment filled tidal channels, raised the elevation of the mudflats and salt marsh, and gradually reduced the tidal prism. With the reduced tidal prism, the effects of freshwater on the normally saline estuary were substantial. Increased freshwater inflows resulted in dramatic shifts in the plant species composition of the salt marsh as well as shifts in the invertebrate and fish communities. Recovery of these communities can take years and was often interrupted by another flood event.

Flooding also dramatically changed the riparian habitats and channels of the Tijuana River. Prior to 1980, the main channel of the Tijuana River ran roughly through the center of the estuary, between the north and south arms. A large pond was created by sand mining activities that existed east of Hollister Street. Following El Niño events in the early 1980s, the river was diverted to the north adjacent to the Naval Outlying Field and the sand mining pit was filled with sediment and the riparian habitats were obliterated by scour (Figure 7). Another storm in 1993 resulted in the river bifurcating at Hollister Street with a new channel created to the north. This phenomenon was attributed to the placement of unpermitted fill within the river channel immediately east and west of Hollister Street Bridge, thereby reducing channel capacity and increasing water velocities.

Historically, flows in the Tijuana River varied from non-existent to torrential floods that inundated much of the river valley before construction of dams began in 1912 (SFEI 2017). The extent of flooding varied depending on the timing and volume of stream flow, the location along the river corridor, and other factors, such as soil moisture. However, during major floods, such as the 75,000 cfs event in 1916, the river could swell to over one kilometer (0.6 mile) in width, filling the valley floor. Portions of floodwaters spilled out of the Tijuana River Valley and flowed north into San Diego Bay on several occasions, including the 1916 event. Such floods scoured and deposited large amounts of sediment, reconfigured river channels and the floodplain, cleared vegetation and damaged infrastructure, such as the railroad, the Tijuana race track and much of the town of Tijuana.

Flooding of Trans-border Canyons. Several tributaries to the Tijuana River flow across the U.S.–Mexico border, and two of these convey substantial sediment loads even during relatively light rainfall events. Smuggler's Gulch crosses the border approximately one mile upstream of the estuary and flows across mostly former agricultural lands before joining the Tijuana River. Goat Canyon crosses the border at the southern arm of the estuary. Both canyons convey flows by means of historically intermittent creeks with their watersheds in Mexico. As presented below, in the 1980s trans-border flows became more pervasive, augmented with untreated sewage from the City of Tijuana.

In addition to Smuggler's Gulch and Goat Canyon, a smaller trans-border canyon,

FIGURE 7. 1980 PHOTOGRAPH OF RIPARIAN HABITATS AT HOLLISTER STREET

known as Yogurt Canyon, also conveys contaminated runoff to the estuary. Yogurt Canyon and associated flows cross the U.S.–Mexico border near Monument Mesa in the far southwest corner of the estuary (Figure 4).

In some years, the City and County of San Diego attempt to remove sediment deposited by Smuggler's Gulch Creek to prevent flooding of Monument Road. The sediment deposited by Goat Canyon Creek entered the estuary by overtopping its banks and sheet flowing into tidal channels and marsh habitats. Thus, Goat Canyon was identified as a major source of sedimentation in the

southern arm of the estuary. The control of this sedimentation is considered a top priority management objective.

Historically, Goat Canyon Creek flowed north across the U.S.–Mexico border, southwest around Spooner's Mesa, then to the west where it often terminated in seasonal wetlands abutting Lichty Mesa (Figure 8). Over the past 30 plus years, unregulated development in the Mexican portion of the watershed has resulted in destabilization of the highly erodible soils of the canyons cut by this creek. As a result, sediment has filled the historic channel of the creek. Prior to construction of the sedimentation

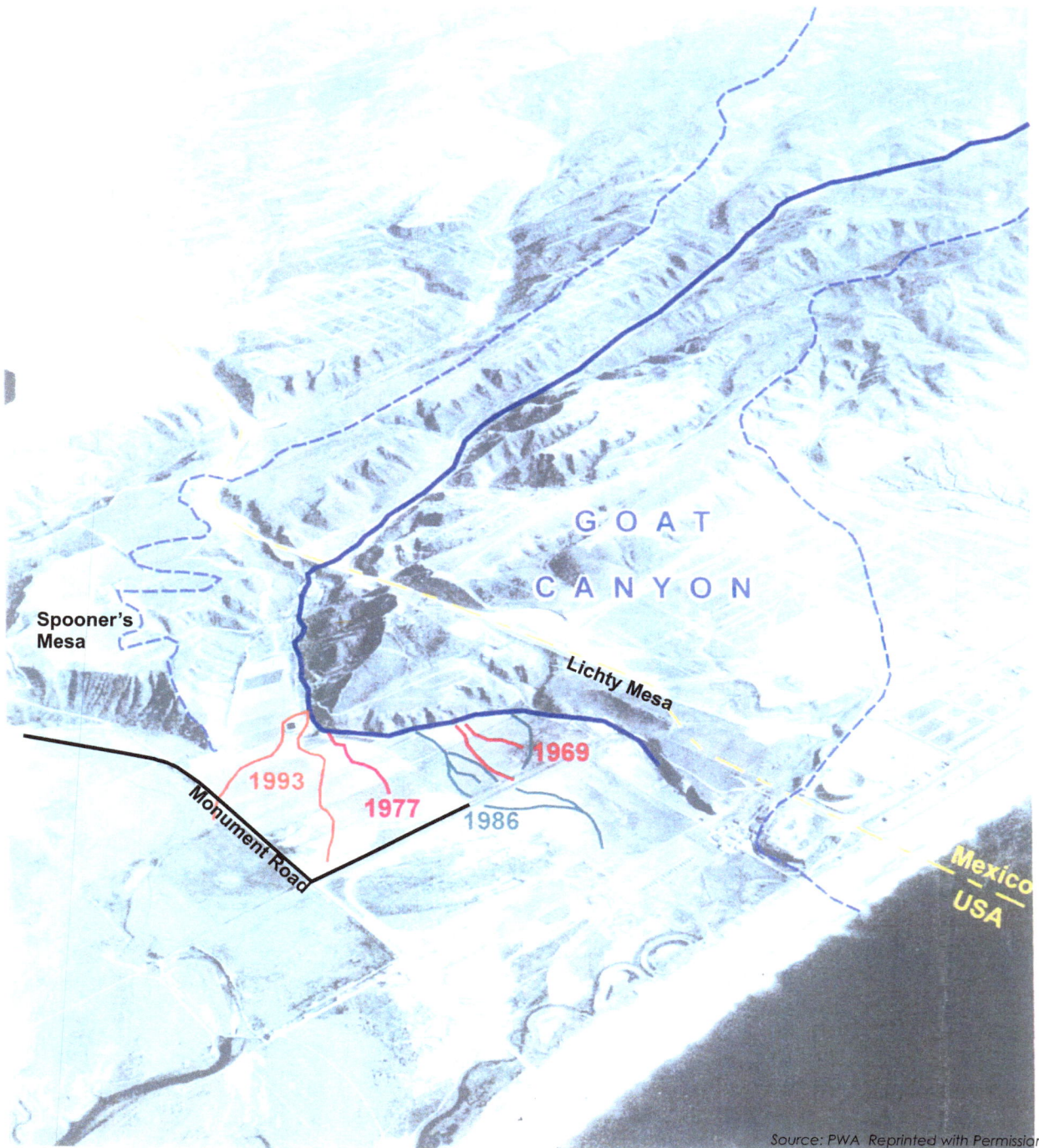

FIGURE 8. HISTORIC MIGRATION OF GOAT CANYON CREEK

basins in 2005, the creek overflowed its channel following only moderate rainfall and deposited sediments over a large area of adjacent uplands and wetlands. In 1969 and 1986, the creek deviated from its historic channel, flowing westward across Monument Road, and then directly into the southern arm of the estuary (PWA 2000). In 1977 and 1993, the creek's course deviated to the northwest, again crossing Monument Road, and entered existing marshland and tidal channels.

Sediment deposited by creek migrations buried Monument Road and have impacted the salt marsh–salt panne habitat west of the road. In addition, seeds of exotic plant species, in particular annual grasses, mustards (*Brassica* spp.) and crown daisies (*Glebionis coronaria*), have invaded the disturbed landscape. As a result, habitat that was once considered valuable seasonal wetland has become both elevated and degraded.

Sediment deposition to the northwest across Monument Road has resulted in similar impacts. The alluvial fan deposited at the end of Goat Canyon has extended across Monument Road resulting in deposition in Old River Slough and associated intertidal wetlands. This deposition has effectively filled the eastern portion of Old River Slough, reducing tidal exchange to a fraction of its former capacity. From 1985 to 1990, the Pacific Estuarine Research Laboratory (PERL) of San Diego State University operated an outdoor laboratory on approximately 70 acres of abandoned agricultural land in the project area. The research laboratory drew water from a shallow pond excavated within Old River Slough. Water was pumped from the shallow pond on high tides and stored in a constructed basin for later use in constructed wetlands and experiments. Today there is no evidence of the shallow pond and that portion of Old River Slough is no longer tidal.

The Goat Canyon sedimentation basins were constructed in 2004–2005 with funding from the State Coastal Conservancy. The basins were designed to capture approximately 60,000 cubic yards of sediment and debris. Due to their relatively small capacity, the basins must be emptied nearly every year. Due to the predominance of fine-grained sediment, disposal of captured sediments can be costly and presents challenges for California State Parks, the agency responsible for management of the basins.

Just prior to the completion of the sediment basins in 2005, a November 2004 storm overwhelmed the basins and up to two feet of sediment was deposited in approximately 20 acres of salt marsh immediately west of Monument Road (Figure 9). This example underscores the magnitude of the sedimentation problems in the south end of the estuary.

Yogurt Canyon continues to impact the southern arm of the estuary. Like the other trans-border canyons, flows have increased in recent years as the population of Tijuana has increased. Seasonally high flows have resulted in the closure of Monument Road

Labels on image: MODEL MARSH, MONUMENT ROAD, SEDIMENT DEPOSITS, GOAT CANYON, RICK ENGINEERING COMPANY

FIGURE 9. SEDIMENT DEPOSITION IN SOUTH ARM

which has restricted access to Monument Mesa in Border Field State Park. Continuous inflows have converted much of the salt marsh in the southwestern portion of the estuary to freshwater marsh with effects on the animal species that rely on salt marsh. The need to control flows from Yogurt Canyon was identified as a priority management objective in the Tijuana Estuary–Friendship Marsh Restoration Feasibility and Design Study (Tierra Environmental Services 2008), a large-scale restoration plan for the south arm of the estuary funded by the Coastal Conservancy.

The historical ecology study found little information on both Goat Canyon and Smuggler's Gulch given their small size and remoteness. Sources indicate that flows through these canyons were intermittent or ephemeral, although Goat Canyon was reported as having an artesian well that provided drinkable water. In some maps, Yogurt Canyon is depicted as salt marsh suggesting that soils were saline and there was little or no freshwater flow.

History of Inlet Closure

Tijuana Estuary has been spared the dissection and filling associated with the construction of highways and the railroad and, thus, has not been subjected to frequent inlet closure as are those lagoons and estuaries that have been impacted by such infrastructure. However, one incident demonstrated the severe effects of such a closure. In 1983, one of the strongest El Niño events on record

affected much of Southern California. Sea temperatures warmed, higher sea levels were recorded (15 centimeters above normal on average; Zedler et al. 1992) and storms were frequent. In January 1983, concurrent high tides and heavy surf resulted in wash-over of the barrier dunes in several locations in the north arm of the estuary. Sand was washed into the main channel of Oneonta Slough reducing the tidal prism and ultimately resulting in closure of the inlet to tidal flushing. Islands that once existed in the main channel were obliterated and the channel was narrowed and constricted.

Inlet closure occurred in early April 1984 after which a dredging plan was developed and implemented by the USFWS. Following a protracted permitting process, the estuary was reopened to tidal flushing in mid-December 1984, eight months after initial closure.

Contrary to 1983, the 1984 rainy season was very dry with almost no rain. Impounded sea water in tidal channels became hypersaline (60 ppt in fall 1984). The water in shallow tidal channels evaporated and the channel bottoms dried and hardened. Marsh soils became desiccated and hypersaline (over 100 ppt in September 1984) and eventually large patches of low salt marsh died. The biodiversity of the estuary was impacted with the loss of native salt marsh species and benthic invertebrate species, some of which were slow to recover (Zedler et al. 1992; Crooks pers. comm). These effects were transferred up the food chain and species such as the endangered light-

footed Ridgway's rail that feed primarily on epibenthic invertebrates disappeared from the estuary. Native dune vegetation that had formerly occurred along the barrier dunes was lost and replaced by non-native species.

The 1984 inlet closure highlighted the fragility of Tijuana Estuary to episodic sedimentation resulting from dune destabilization and overwash. Researchers from SDSU and TRNERR undertook dune revegetation programs and long-term monitoring of salt marsh, fishes, invertebrates and birds was initiated immediately after tidal flushing was restored. The monitoring program revealed that some species had recovered from the inlet closure but that this recovery was hindered by shifts in estuarine water salinities from increased sewage flows in the mid- to late 1980s.

The inlet has mostly remained open since the 1984 closure; however, reduction of tidal prism in the north arm from dune overwash coupled with a similar loss of prism in the south arm from sediment deposited from trans-border canyons has resulted in a tenuous situation where closure is more probable. During winter of 2009, the inlet began migrating south of its typical position in the mid-estuary. Beach sand was deposited by storm waves at the confluence of the Tijuana River and the south arm, temporarily blocking most tidal flows to the south arm. The USFWS funded removal of the sand which was placed on the adjacent dunes. This stop-gap measure succeeded in restoring tidal flows to the south arm but

demonstrates the sensitivity of the system to additional closures. More recently, the inlet has closed or partially closed on at least five occasions between 2016 and 2017 The USFWS removed sand from the inlet on each occasion, depositing the sand on the dunes south of the inlet. Closure in March 2016, coupled with sewage-contaminated water in the Tijuana River, resulted in the die-off of fish and invertebrates, including numerous leopard sharks. The long-term trajectory of the inlet following these closures is not known at this time.

HISTORY OF INLET AND CHANNEL MAINTENANCE

In response to the 1984 inlet closure, the USFWS dredged the main channel of Oneonta Slough from the end of Seacoast Drive to the inlet with a clam shell dredge. The sand dredged from the channel was side cast and bulldozed to recreate the dunes which became part of a continuing dune management program in the northern arm. Additional dredging in the south arm was conducted in 1986 with similar dune restoration.

The more recent blockages were also removed by the USFWS. Sand that was washed into the main channel connecting the inlet to the south arm was excavated with land-based equipment (excavators and front loaders) and deposited in the dunes above the highest tides. There was no attempt to establish dune vegetation on the deposited sand.

History of Wastewater Inflows

Historically, the Tijuana River typically had little or no flow in summer months when rainfall is low and evaporation rates are high (Zedler et al. 1992; Nordby and Zedler 1991; SFEI 2017). For over 70 years, the Tijuana River has received raw sewage flows from the City of Tijuana increasing in volume to an estimated average of 10–12 million gallons per day (mgd) in the late 1980s. Renegade flows were estimated at 22 mgd in 1987–1988 (Seamans 1988). Intermittent sewage flows also entered the estuary from Goat Canyon and Smuggler's Gulch. Smuggler's Gulch conveyed an estimated four–five mgd of sewage to the estuary during the same period These combined flows affected the species composition of the salt marsh plant community as well as the communities of benthic invertebrates and fishes.

In 1988, the IBWC built an interceptor to collect and return Smuggler's Gulch sewage flows and return them to the Tijuana treatment system. During the mid-1990s, the City of Tijuana upgraded its sewage system to increase its capacity and improve conveyance pipelines. In 1997 the IBWC built the International Wastewater Treatment Plant (IWTP) in the Tijuana River Valley and two years later completed the ocean discharge pipeline (South Bay Ocean Outfall). With the completion of the IWTP, the interceptor in Smuggler's Gulch and a later interceptor built in Goat Canyon, sewage flows were piped directly to the IWTP rather than returned to the Tijuana

treatment system. Designed to capture and treat 25,000 mgd of sewage from Tijuana, the plant can be overwhelmed by volumes that exceed 25,000 mgd which are then released to the Tijuana River and estuary.

During the peak flows in the late 1980s, wastewater inflows from Tijuana had dramatic effects on the biota of the estuary. Although these flows were contaminated with sewage, it was concluded that the reduced salinities associated with the increased input of contaminated freshwater was the primary cause of these impacts (Nordby and Zedler 1991). The structure of the fish assemblage shifted toward dominance by species with extended spawning seasons and rapid maturity. Bivalve populations were dominated by young individuals as most older individuals died off. Polychaete populations were dominated by taxa associated with pollution that had extended spawning seasons and rapid maturity.

Like the effects of flooding and sedimentation, recovery from the effects of wastewater on the biota of the estuary was dependent on the elimination of the source of disturbance. Long-term monitoring of the estuary has documented the reversal of many of these impacts.

Impacts Associated with Construction of the New Border Fence

In 1996, the U. S. Congress approved the construction of a new border fence along the

U.S.–Mexico border, beginning at the Pacific Ocean at Tijuana Estuary and extending 14 miles eastward in San Diego County. The new fence was to be approximately three times the width of the existing fence, consisting of three parallel 10- to 15-foot-high steel walls, with 50 feet of land between each wall graded and cleared of all vegetation, and the entire area lit by floodlights. The U.S. Customs and Border Protection also proposed filling in canyons and leveling existing uplands in order to provide a roadway that could accommodate high speed vehicles. In 2004, the California Coastal Commission determined that the border fence project would violate the Coastal Zone Management Act and damage the valuable resources of the Tijuana River National Estuarine Research Reserve (TRNERR). The Sierra Club, Audubon Society, and other environmental groups also challenged the border fence in court and construction was halted.

In response, Congress passed the Real ID Act of 2005 which afforded the Secretary of the Department of Homeland Security (DHS) the power to waive all laws that might slow or stop construction of the border fence. Homeland Security subsequently waived in their entirety the Coastal Zone Management Act, the National Environmental Policy Act, the Endangered Species Act, the Migratory Bird Treaty Act, the Clean Water Act, the Clean Air Act, and the National Historic Preservation Act in order to continue construction. Accordingly, the DHS filled in Smuggler's Gulch with over 2,000,000 cubic yards (yd^3) of earth that had been excavated from adjacent sensitive upland habitats within the TRNERR and County Park, and constructed the border fence on top of the berm (Figure 10). There was little or no effort to revegetate the bare slopes which threatened to wash into the estuary. Eventually, with pressure from resource managers, a revegetation program was initiated along some of the slopes.

The 2004 lawsuit filed against the DHS by the State of California and environmental groups representing CSP was settled out of court. Approximately $5,000,000 was allocated to CSP for restitution for the damages, which is being used to plan for improved access to Border Field.

IMPACTS ASSOCIATED WITH INSECT-BORNE PATHOGENS

In 2015, managers within the Tijuana River Valley began noticing die-off of the willow woodland associated with the river. After close analysis, it was determined that the cause was two species of non-native beetle, the Kuroshio shot hole borer beetle. Both are members of the genus *Euwallacea*. These beetles burrow into tree bark and infect the trees with at least two fungal species. The fungus grows within the tree and the beetle larvae feed on the fungus until maturity. The fungus disrupts the tree's metabolic processes in a process known as "Fusarium die-back" (*Fusarium euwallaceae*) leading to branch loss and ultimately, in many cases, death of the trees. Willows are not the only type of trees affected. Other species include sycamores and oaks, avocado and

FIGURE 10. SMUGGLER'S GULCH FILL

other commercial and ornamental varieties. Thus, there is the potential for significant economic loss in southern California.

Fusarium die-back has devastated the willow woodland in the river valley. A recent estimate (January 2016) concluded that over 130,000 native riparian trees have already been affected. This number includes trees that are already dead or are significantly infested with the beetles. Even though the growth of riparian in the valley has been fueled by cross-border flows of water, extant habitat is viewed as an important resource given impacts to coastal riparian zones throughout the region. Many of the riparian restoration projects presented in this document have been affected and the once-vibrant riparian habitats are now primarily dead (Figure 11). It is not known how this will affect breeding by song birds in the valley, particularly the endangered least Bell's vireo (*Vireo bellii pusillus*) which nests in dense will scrub.

There are no effective control measures for this pest complex presently identified. Single

FIGURE 11. SHOT HOLE BORER DEVASTATION

tree treatments with various pesticides have been shown to be effective but the sheer number and diversity of tree species affected makes spot treatment an unrealistic means for controlling infestations. A wasp parasite of the beetles has been found in Asia but has not yet been tested for its potential safety to be introduced into North America as a controlling predator of these beetles. Researchers from the University of California Riverside, U.S. Forest Service and others are leading the investigations into this emergent pest. The California Avocado Growers Association has funded the bulk of scientific investigations addressing this problem to date.

Should a remedy be identified or a resistant strain of willow discovered, substantial restoration efforts in the valley may be warranted. This phenomenon is not unique to the Tijuana River Valley and the beetles have been reported from other Southern California watersheds.

According to Dr. John Boland, an ecologist who has worked extensively in the Tijuana River Valley, the shot hole borer infestation was unique in the river valley compared to other southern California riparian systems. He hypothesized that the sewage that continues to contaminate the Tijuana River acts as fertilizer that accelerates growth of willows. Such accelerated growth results in soft tissue with high water content that is favored by the shot hole borer relative to drier areas where growth is slower and the wood harder with lower water content (Boland 2018). To test this "Soft Tree" hypothesis, he compared tree height, wood density, water content and shot hole borer infestation rates of willow stands near and far from the flows of the Tijuana River. The results of these comparisons supported the hypothesis. Trees within the Tijuana River Valley subjected to the sewage-rich river flows were significantly taller than the same-aged trees growing outside the valley. Wood samples from willow trees had significantly lower densities and higher water contents at sites near the Tijuana River flow, indicating that the Tijuana River trees were 'softer' in comparison to sites far from the polluted flows.

Contrary to initial assessments of mortality, Dr. Boland finds that the heavily infested willow forest of the Tijuana River Valley is recovering. Now many willows damaged by the shot hole borer have re-sprouted vigorously from their stumps, and seedlings have established in large patches. By late 2017, the developing canopy was about five meters tall. Infestation rates decreased to only 6%, down dramatically from its initial 97% in 2015–2016.

TIMELINE OF RESTORATION PROJECTS AT TIJUANA ESTUARY

ALTHOUGH MANY of the restoration efforts in the river valley and the estuary were responses to the impacts previously presented, the earliest projects were efforts to understand restoration fundamentals, such as how to propagate and install plants, how to improve drainage in the root zone, the effects of restoring tidal flushing to areas that had been bermed or diked to prevent such flushing and restoration of the barrier dunes.

This work began in the late 1970s and early 1980s (Table 1) primarily through research by Dr. Joy Zedler, graduate students, and postdoctoral researchers at SDSU. With the formation of the Pacific Estuarine Research Laboratory (PERL) in 1985, research on restoration expanded to an ecosystem level approach that included nutrient cycling, algal productivity, competition between various plant species, the effects of renegade sewage on aquatic organisms and other topics.

In 1989, Dr. Zedler and associates along with hydrologists from the firm Phillip Williams and Associates (PWA), developed an ambitious plan to restore approximately 495 acres of wetland habitat in the south arm of the estuary, funded by the State Coastal Conservancy. The product, the Tijuana Estuary Tidal Restoration Program (TETRP), was presented in an Environmental Impact Report (EIR)/Environmental Impact Statement (EIS) in 1991.

TABLE 1. TIMELINE OF RESTORATION ACTIVITIES AT TIJUANA ESTUARY

Project Number	Date	Project	Responsible Party(ies)
1	1976	Clapper Rail Habitat: Requirements and Improvement	Zedler et al.
2	1980	Coastal Wetlands Restoration and Enhancement	Zedler et al.
3	1980	Coastal Wetlands Management: Restoration and Enhancement	Zedler et al.
4	1985	Dune Restoration at Tijuana Estuary	CSP, USFWS, PERL
5	1987	Dune Vegetation Reestablishment	Zedler and Woods
6	1989	Border Field State Park Dune Restoration Project	CSP, PERL
7	1991–2004	Tijuana Estuary Tidal Restoration Plan (TETRP)	PERL/PWA/ENTRIX
8	2003–2008	Tijuana Estuary–Friendship Marsh Restoration Project—Feasibility and Design Study	Coastal Conservancy/SWIA/Nordby et al.
9	1992–2002	Restoration of Former Model Airplane Field	CSP
10	1994–2012	FUDS MMRP Project, Border Field State Park Remediation Investigation/Feasibility Study	U.S. Army Corps of Engineers/CSP/USFWS/California Department of Toxic Substance Control/Bristol Environmental Remediation Services
11	1996–2006	South Bay Water Reclamation Plant and Dairy Mart Road Improvement Project	City of San Diego/Nordby/Mooney & Associates/Merkel & Associates
12	1997–2002	Hollister Street Bridge Replacement Project	City of San Diego/Nordby
13	1998–2001	Napolitano Restoration Project	Caltrans District 11/Nordby
14	1998–2009	Goat Canyon Enhancement Project	CSP/Coastal Conservancy/SWIA/Nordby/Burkhart/EDAW
15	2001–2006	Fenton Quarry Restoration	Coastal Conservancy/SWIA/Nordby
16	2002–2018	Tijuana Valley Invasive Plant Control Project	Coastal Conservancy/USFWS/RWQCB/SWIA/Boland/City of San Diego
17	2004–2010	Tijuana River Valley Regional Park Trails and Habitat Enhancement Project	County of San Diego/Greystone Environmental Consultants/Kimley-Horn/Rick Engineering/Nordby
18	2008–2009	Tijuana Estuary Fate and Transport Project	CSP/Coastal Conservancy/USGS/SWIA/TRNERR
19	2009–2011	WRT Transition Zone Restoration	Talley et al.
20	2009–2014	Tijuana River Estuary Water Quality Improvement and Community Outreach Project	TRNERR/CSP/SWIA
21	2011–2013	Nelson Sloan Quarry Restoration Plan	Coastal Conservancy/City of San Diego
22	2011–present	Five-Acre Restoration Project in Border Field State Park	IBWC/URS/SWIA/ECM/ASCE/Boland
23	2012–present	Bunker Hill Revegetation: Border Infrastructure System	U.S. Customs and Border Protection/HDR/RECON

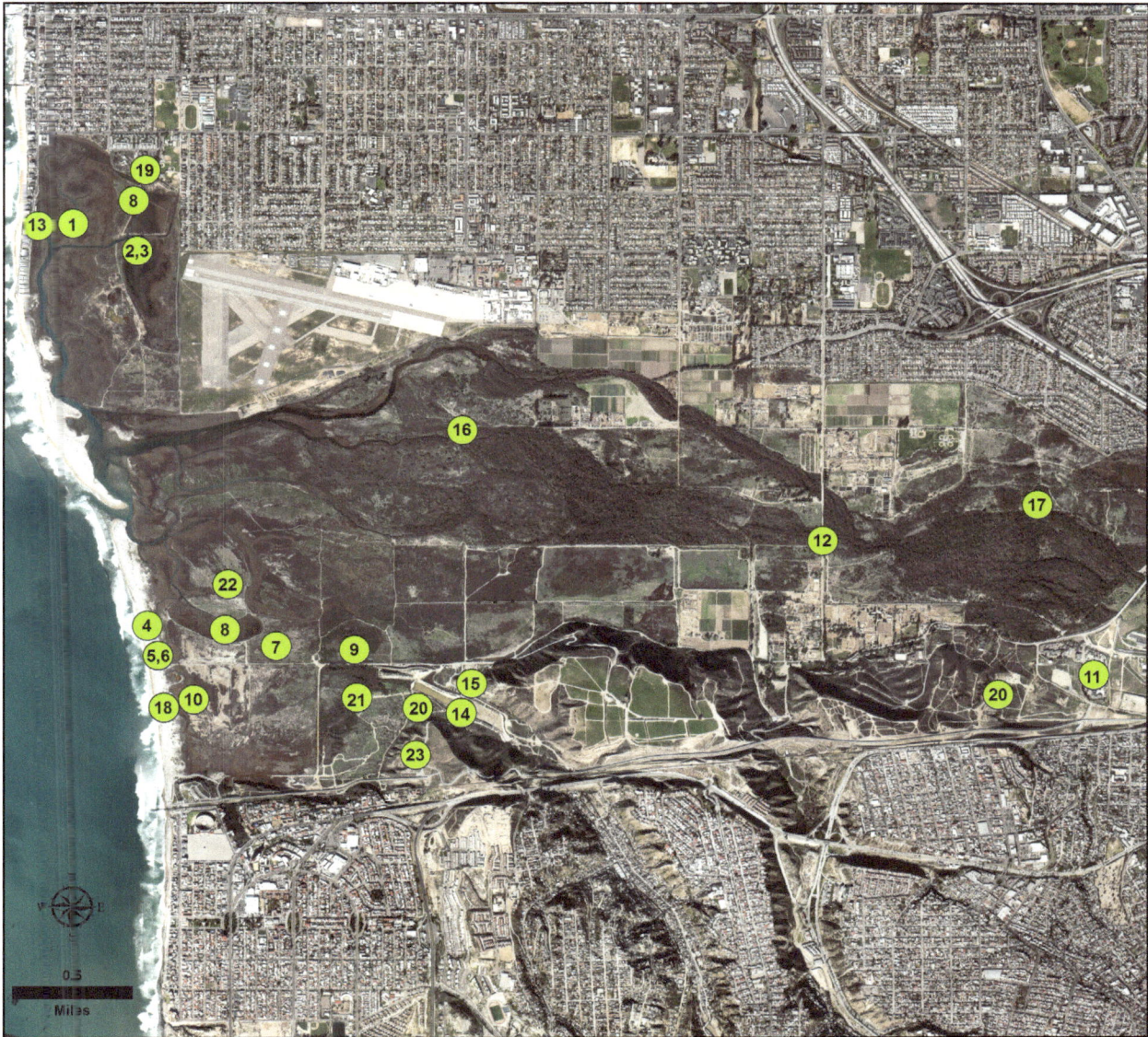

FIGURE 12. LOCATIONS OF RESTORATION PROJECTS IN THE VALLEY

One controversial aspect of the project, river training berms 12–25 feet in height and 12–46 acres in area (depending on alternative), generated concern by state and federal regulatory agencies and the 495-acre restoration was never constructed.

Two pilot projects identified in TETRP were subsequently implemented: the approximately two-acre Oneonta Slough Connector Channel (also known as the Tidal Linkage) in 1996–1997 and the 20-acre Model Marsh (also known as Friendship Marsh) in 1999–2000. Both incorporated restoration research and monitoring to determine achievement of success criteria. Sediment excavated to restore the Model Marsh was subsequently used to restore an abandoned sand and gravel quarry in Goat Canyon with maritime succulent scrub habitat in 2001.

In 1998, Caltrans District 11 restored a 1.25-acre parcel of former wetland at Tijuana Estuary that had been filled and was being used as a parking lot by area residents. This parcel, known as the Napolitano site, was formerly privately owned. The restoration was conducted under a Supplemental Environmental Project (SEP) as partial mitigation for storm drain discharge to San Diego Bay as a consequence of a lawsuit brought against Caltrans by local environmental groups.

In 2004–2005, the Goat Canyon Sediment Basins were constructed in response to the sedimentation issues plaguing the estuary. The sediment basins not only were successful in capturing and removing sediment before it could reach the estuary but also included restoration of approximately 20 acres of mule-fat scrub and 1.6 acres of southern willow scrub as well as upland habitats as mitigation for project impacts.

A second restoration plan for the south arm of Tijuana Estuary was completed in 2008. This project, known as the Tijuana Estuary–Friendship Marsh Feasibility and Design Study, identified approximately 250 acres of restoration that if implemented would nearly double the tidal prism of the estuary.

In 2008 and 2009, the USGS and State Coastal Conservancy teamed to conduct a test of the environmental effects of disposing fine sediments in the nearshore environment. Approximately 45,000 yd^3 of sediment was trucked from the Goat Canyon stockpile and deposited in the surf zone. The USGS tracked the fate of the sediment while biologists contracted by SWIA and from Scripps Institution of Oceanography studied the effects on a wide variety of organisms. In 2009, restoration of upland and transitional habitats near the Tijuana Estuary visitor center was undertaken by students of Dr. Theresa Talley of the University of San Diego/ California Sea Grant Scripps Institution of Oceanography. This restoration was small in scale (<0.5 acre) and was experimental in nature. The site was part of a former landfill and was significantly degraded. This project complemented on-going efforts by CSP to remove monotypic stands of the native big saltbush (*Atriplex lentiformis*) from the grounds of the visitor center and replace them with a more diverse assemblage of native upland species.

Removal of exotic invasive species from riparian and upland habitats in the river valley has been conducted with numerous granting agencies from 2002 through 2018. This work, conducted by Dr. John Boland of Boland Ecological Services and subcontractors administered by SWIA, reduced the population of target invasive species substantially and facilitated both natural and active revegetation by native species.

The IBWC began restoration of a five-acre site in Border Field State Park in 2011 as partial mitigation for the loss of raptor foraging habitat associated with upgrades to the South Bay International Wastewater Treatment Plant.

The U.S. Customs and Border Protection recently initiated vegetating the exposed areas associated with the Border Infrastructure System, or border fence. Invasive non-natives have been treated and removed from the site and native species have been planted and are currently being maintained and monitored.

In May 2016, the State of California Wildlife Conservation Board funded the first phase of the Tijuana Estuary Tidal Restoration Program in the south arm of the estuary, primarily within Border Field State Park. The project is based on the work completed in 2008 (Tijuana Estuary–Friendship Marsh Restoration Project Feasibility and Design Study). The goal is to restore approximately 75 acres of upland and degraded wetland.

Each of the projects described here, as well as other studies and on-going research on restoration at Tijuana Estuary, are described in detail in the following sections. A map depicting the locations of each of the restoration sites listed in Table 1 is presented in Figure 12.

Many valuable lessons were learned from their planning, permitting and implementation that are applicable to future restoration efforts. Although intended to be a complete accounting of all restoration projects in the river valley, certain entities would not or were unable to provide documentation of past restoration efforts. Nonetheless, the majority of such efforts are represented in this document.

Restoration in the Tijuana River Valley and estuary is unique among southern California coastal wetlands in that most of the area occurs within the boundaries of the TRNERR administered by NOAA. As a research reserve, the TRNERR facilitates the incorporation of scientific research to inform restoration techniques and processes. One of the major contributions to restoration ecology employed by the TRNERR and Dr. Zedler and her colleagues is the application of adaptive management, wherein experiments were incorporated into restoration projects to inform future projects. Adaptive restoration was a prominent feature of both the Oneonta Slough Connector Channel (Tidal Linkage) and Model Marsh (Friendship Marsh) restorations, as demonstrated in subsequent sections of this report.

It should be noted here that there is an extensive body of research on salt marsh ecology, including research on salt marsh restoration, at Tijuana Estuary and other regional wetlands that overlaps the projects presented in the following sections. The focus of this document is on the lessons learned from restoration projects that were implemented and monitored in the Tijuana River Valley and Tijuana Estuary. The need for an annotated bibliography of all restoration projects and research in the valley and estuary has been identified. It is anticipated that such a bibliography will be produced as a follow up to this document.

RESTORATION EFFORTS ASSOCIATED WITH SAN DIEGO STATE UNIVERSITY

1 CLAPPER RAIL HABITAT: REQUIREMENTS AND IMPROVEMENT

Principal Investigator:
    Joy B. Zedler, San Diego State University, Department of Biology

Research Assistants:
    Chris Nordby, San Diego State University, Department of Biology
    Phil Williams, San Diego State University, Department of Biology

Date: 1976

Funding Agency:
    U.S. Fish and Wildlife Service

Reported in:
    Zedler, J.B., C. S. Nordby and P. Williams. Clapper Rail Habitat: Requirements
    and Improvement. Final Report to U.S. Department of Interior, Fish and
    Wildlife Service. December 1979.

The objectives of this research were twofold:

1) to determine what factors controlled the distribution of California cordgrass (*Spartina foliosa*) and, thus, the distribution of light-footed Ridgway's rail, at Tijuana Estuary; and,
2) to determine how cordgrass establishes naturally and how it can be cultivated or transplanted. It was during this time period that permanent transects were established in the salt marsh of the north arm of the estuary that served as the beginning of a long-term monitoring program.

Prior to conducting research of these stated objectives, a census of nesting clapper rails in the north arm of the estuary was conducted by Paul Jorgensen of the U.S. Navy, Jim Zimmer of SDSU, and Greg Loeb, SDSU graduate student. The surveyors repeated Paul Jorgensen's clapper rail surveys conducted in 1974 when he reported 22 nests in cordgrass-dominated salt marsh in Oneonta Slough. The 1979 survey located 16 nests, all of which but five were located with cordgrass present or dominant, thus supporting Jorgensen's conclusions that cordgrass is the preferred nesting habitat for this species.

Factors Controlling the Distribution of Cordgrass. This portion of the study examined five primary research questions:

- What are the characteristics of the low marsh vegetation and environment at Tijuana Estuary?

- What factors control cordgrass growth and reproduction?
- What factors explain the variability in height, density, and flowering of cordgrass at Tijuana Estuary?
- Is the population of cordgrass at Tijuana Estuary expanding, receding or remaining stable?
- Is cordgrass an invader of disturbed habitats?

METHODS

In order to characterize the environment in which cordgrass grows at Tijuana Estuary, Zedler and associates established transects across its distribution at eight locations (Figure 13). The transects were located to include the full range of variability in cordgrass height and density at the estuary. The beginning of each transect was located randomly within the chosen sampling area, and permanent stakes were placed at five m intervals from the seaward to landward boundaries of cordgrass distribution. In all, 101 sampling stations were established and the following features of the vegetation and environment were measured at each:

- Height and density of cordgrass within 0.25 square meter (m^2) circular quadrats at the end of the 1979 growing season (August and September);
- Elevation;
- Interstitial soil salinity;
- Soil oxygen concentration;
- Soil moisture;
- Soil bulk density;
- Soil organic matter;

- Soil nitrogen;
- Soil water temperature; and,
- Cover of other species present.

By examining these variables, the researchers hoped to determine cause–effect relationships through multiple regression analysis.

RESULTS

The 101 sampling stations included 19 that were located in monotypic cordgrass; 73 that were a mixture of cordgrass and other species; and nine stations that did not contain cordgrass. In contrast to the general description of salt marsh elevation zonation, there was no clear relationship between cordgrass and absolute elevation when examined in detail. Cordgrass distribution was not limited to the lowest elevations nor was it always found within the same elevation band. At one transect (TJE-37), cordgrass was restricted to stations below about +70 centimeters mean sea level (MSL) while at another (TJE-5), it occurred as high as +85 centimeters MSL and as a monotype at elevations over +75 centimeters MSL.

It had been hypothesized that the reduction in clapper rail nests seen in 1979 (16 nests) compared to 1974 (22 nests) had been the result of a decline in formerly pure stands of cordgrass and replacement with a mixture of cordgrass and succulent species. In order to test this hypothesis, eight areas of cordgrass within Oneonta Slough that were thought to be declining at their landward boundary were selected for a detailed study of height,

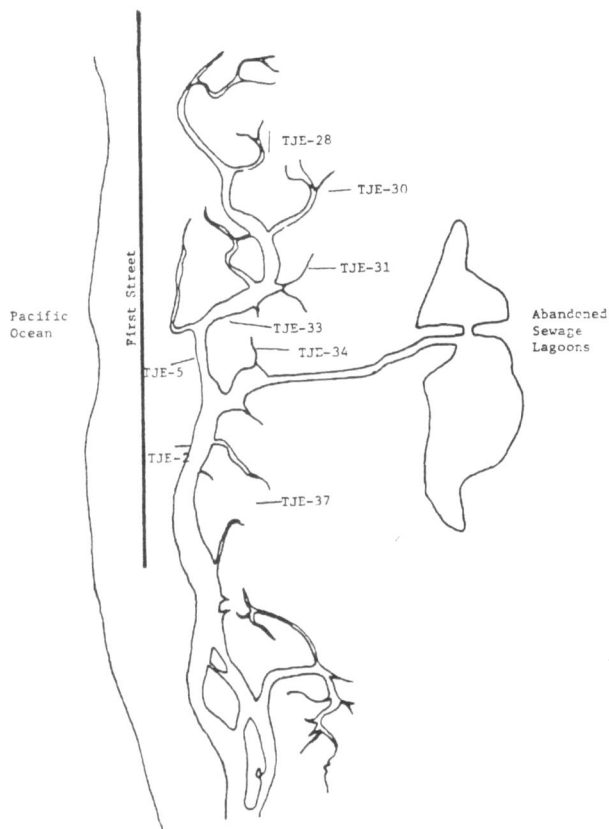

FIGURE 13.
CORDGRASS TRANSECTS ESTABLISHED
IN 1980

density and number of live and dead stems. The ratio of live to dead stems was used to determine whether that particular stand was declining or increasing (Zedler et al. 1979).

In general, the physical environment within cordgrass distribution at Tijuana Estuary is low in elevation (mean = +67 cm MSL); moist (mean = 250% in soil); hypersaline in fall (mean = 44 ppt); nearly anaerobic (< 2 ppm O_2); and highly organic (mean = 20% in soil). Soils have relatively low density, due to the presence of water-logged clays and organic matter. However, all of these environmental factors were highly variable among the sampling stations with standard deviations ranging from 11% to 64% of the mean.

The relationship of environmental factors and vegetation to one another using correlation analysis revealed several significant correlations among the 11 factors measured. The correlations with elevation led to the following general hypothesis: where elevations are high, organic matter is high, which leads to low bulk density (because of the low weight of the organic matter); high soil moisture in turn leads to low soil temperature.

Soil salinity showed few correlations in the 92 stations where cordgrass occurred. Within this zone, salinity is not related to elevation but does correlate with soil moisture, suggesting that evaporation increases soil salinity in the low marsh. At the larger scale, soil salinity generally decreased with elevation. The salinity environment was thus described as increasing from the lower elevations where cordgrass is found (edges of tidal channels) to the upper limit of cordgrass, and then decreases toward the upper intertidal boundary.

At its seaward limit, cordgrass exists at elevations as low as MSL at Tijuana Estuary. As vascular plants are essentially land plants which have invaded the intertidal environment, the authors postulated that it was reasonable to assume that their seaward limit would be attributed to the inability to tolerate frequent long-term inundation. However, the lower limit of cordgrass was not investigated in the study.

The general conclusions from the multiple regression analysis were: there is little relationship between the measured environmental variables and cordgrass vigor within its zone of occurrence at Tijuana Estuary, even where it grows monotypically; interspecific competition is probably important in determining height, density and flowering; and, although elevation is a good predictor of optimum cordgrass habitat on the whole-marsh scale, it is less useful on the small scale. The location of a high or low point at some given elevation relative to tidal channels and surrounding microtopography is more important than absolute elevation. Each of these factors informs the design of restoration projects. Where the goal is restoring clapper rail habitat, understanding the factors that influence cordgrass growth is vital.

The analysis of whether cordgrass was expanding or declining at Tijuana Estuary suggested that five of the eight areas were expanding, one was stable and two were declining. Thus, the hypothesis that cordgrass was declining was not supported.

The study concluded that cordgrass was a poor invader of disturbed habitats as the percentage of seed germination in the laboratory is low and, as of the date of this study, seedlings had not been observed in the natural marsh. As stands of cordgrass increase their area by vegetative spread, the origins of existing populations are unclear. It was postulated that dispersal by clumps of plants being dislodged and floating to a new site was possible. Patches of cordgrass that were disjunct from the nearby stands were observed at the former sewage lagoons at Tijuana Estuary. Lacking the knowledge of past disturbance, it was not possible to determine whether these were the result of recent invasion or remnant patches of former cordgrass stands.*

Twenty-four patches of cordgrass located along the margin of the lagoons were studied to document the rates of vegetative expansion. The patches were staked to mark their north–south and east–west boundaries and the two diameters were measured on March 1979 and at the end of the growing season (October 1979). The average increase in diameter was 0.94 meter (Zedler et al. 1979).

To determine the most effective method for establishing cordgrass for the purposes of restoration, three transplant methods were compared. In decreasing order of likelihood of success these were entire plants with soil cores attached (also known as "plugs"), entire plants with root mass only (no soil), and rhizome segments with nodes and tips.

It was determined that plugs were the most resistant to environmental stress and further analysis of this technique was undertaken. It was documented through experimental transplantation that plugs can be relocated at any time of year. Salinity shock was suggested as the main source of mortality and it was recommended that transplantation into areas with a salinity above 50 parts per thousand (ppt) be avoided. Herbivory by small mammals (rabbits and rodents) was a major source of mortality and experimental plantings required aviary wire cages to exclude herbivores. Future experiments examining the interactions between cordgrass and pickleweed were planned in an attempt to determine the effect of interspecific competition.

LESSONS LEARNED

The results of this study imparted important and specific lessons learned for future restoration efforts. The primary successful method for introducing cordgrass to restoration sites remains plants with native soil intact and plants with bare roots. Plants with bare roots experience much greater transplantation shock, but often enough survive to expand vegetatively. It is not possible to control environmental factors

It should be noted that the conclusion that cordgrass was a poor invader was later negated by observations of natural recruitment in the former sewage lagoons located in the north arm of Tijuana Estuary (Ward et al. 2003). Following El Niño generated winter storms in 1993, cordgrass clones established in the 16-acre mudflat with more than 80 new clones counted by 1998. These storms apparently facilitated habitat conversion through river flooding which caused a temporary reduction in soil salinity and delivered large volumes of sediment that likely raised mudflat elevations enough to allow cordgrass establishment. Today the lagoons are densely populated by cordgrass.

at restoration sites and cordgrass habitat is typically based on an elevational range similar to those measured by this study. That range is typically based on inundation frequencies with cordgrass occurring at elevations that are in general inundated 25%–45% of the time, depending upon variations in tidal amplitude at various locations. Restoration of low salt marsh dominated by cordgrass remains a high priority management objective for many marsh restoration projects for the recovery of the endangered light-footed Ridgway's rail.

2 COASTAL WETLANDS RESTORATION AND ENHANCEMENT

Principal Investigator:
   Joy B. Zedler, San Diego State University, Department of Biology
Research Assistants:
   Chris Nordby, San Diego State University, Department of Biology
   Phil Williams, San Diego State University, Department of Biology
   John Boland, San Diego State University, Department of Biology
Date: 1980
Funding Agency:
   U.S. Department of the Navy
Reported in:
   Nordby, C., J. Zedler, P. Williams and J. Boland. Coastal Wetlands Restoration and Enhancement. Final Report to the U.S. Department of the Navy. November 1980.

This project was a follow-up to the 1979 study and focused on methods for restoring and enhancing salt marsh vegetation in disturbed areas. It was conducted at one of the abandoned sewage oxidation lagoons at Tijuana Estuary. It focused on the low elevation salt marsh dominated by cordgrass as this community is the least likely to reestablish after disturbance. The work involved:

- Planting cordgrass seedlings and plugs in the disturbed sewage lagoon to expand low marsh habitat;
- Monitoring natural expansion of established cordgrass patches at the sewage lagoon;
- Manipulating existing patches of cordgrass by adding nitrogen and tiles to increase surface drainage;
- Treating transplanted "gardens" of cordgrass with nitrogen and tiling to improve drainage;

- Determining the ability of cordgrass seedlings grown in fresh, brackish and sea water to withstand transplant shock.

METHODS

Seedling Transplants. Two seedling transplant gardens were established on the west side of the southern sewage lagoon in an area that was originally lacking cordgrass (Figure 14). In the first transplant garden (garden #1), 66 cores containing from one to five aerial shoots with rhizomes and native soil were transplanted in 1979. An additional 20 ramets of cordgrass propagated from seed in the laboratory were transplanted in August 1980.

A second seedling transplant garden (garden #2) was established in October 1980 to test for differences in transplant shock between seedlings grown in fresh, brackish and sea water. Twenty-nine ramets grown from seed were selected based on similar size and age. These were planted at one m intervals south of transplant garden #1. Both transplant gardens were protected with one-foot high aviary wire to prevent grazing which was a cause of high mortality in earlier transplant experiments.

Transplant Gardens from Mature Stands. One hundred shoots of cordgrass were transplanted along a tidal creek located at the south end of the sewage lagoon in June 1980 (Figure 14). The creek banks appeared to be suitable habitat for cordgrass, yet none had become established.

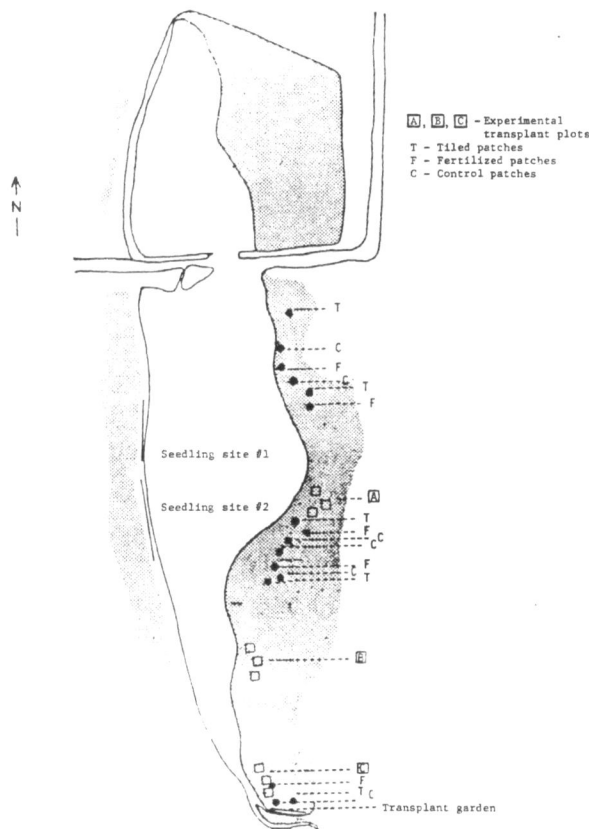

FIGURE 14.
CORDGRASS SEEDLING TRANSPLANT GARDENS 1980

Nine experimental transplant gardens were established in June 1980 to test the effects of tiling to improve surface drainage and fertilization with nitrogen in the form of urea. Three groups of three, two-m^2 plots were established in locations A, B and C (Figure 14). Thirty-six cordgrass sprigs were transplanted in each plot. Within each group, one plot was fertilized with 10 gallons of urea/m^2 every two weeks, one plot was tiled with Spanish-type roofing tiles to improve surface drainage, and one plot served as a control.

Together, a total of 840 shoots and seedlings with a total area of 138 m² were transplanted. In general, the high success rate of establishing cordgrass outside the range of existing stands strongly supported the hypothesis that this species is limited by dispersal.

Growth Experiments on Naturally Established Cordgrass Patches. Fifteen cordgrass patches of similar size were assigned treatment similar to the nine transplanted patches: Five were used as controls, five were treated with 10 gallons urea/m² every two weeks and five were tiled. Initial heights and densities of cordgrass were assessed in June 1980 within two randomly selected 0.25 m² circular quadrats/patch. In September 1980, the plots were recensused and the differences in growth by treatment were analyzed using two-way ANOVA.

RESULTS

Seedling Transplants. Of the 66 cores transplanted to garden #1 in 1979, only 18 survived. High soils salinities were postulated as the cause of transplant shock. By comparison, the subsequent planting of 20 clumps of cordgrass raised from seed in August 1980 had 100% survival and measurable new growth. All but three increased vegetatively, forming new shoots. This difference in survival was attributed to decreased interstitial soil salinities from winter flooding in 1980. This is discussed in greater detail under Monitoring of Natural Cordgrass Expansion.

Transplant Gardens from Mature Stands. The results of experiment on the effects of tiling and nitrogen addition on cordgrass growth in transplanted gardens were deferred to the next growing season and conducted under a different grant. These are presented in the Coastal Wetlands Management: Restoration and Enhancement project.

Growth Experiments on Naturally Established Cordgrass Patches. ANOVA results indicated that tiling to improve surface drainage and addition of urea did not significantly increase cordgrass growth relative to control plots. These results were interpreted in several ways:

1) The timing of the experiment made it difficult to obtain an effect. Since the experiment was begun late in the growing season, the treatments may have occurred after the exponential growth phase of cordgrass, and thus had an imperceptible effect on the plants or the measurement period (June–August) may have been too brief to see any long-term response.

2) Nitrogen may not have been a limiting resource in 1980 due to accumulation of nutrient-rich soils by flooding from winter storms. Additionally, it was possible soil nitrogen was already high at the sewage lagoons due to their former use.

3) While tiling accelerated surface drainage, the effect on the root zone may have been insufficient to produce a measurable effect on plant growth. However, installing tiles below the root

zone would have required additional digging and associated habitat impacts. Several patches of cordgrass showed greater expansion into pools of standing water than into better drained sites.

Monitoring of Natural Cordgrass Expansion. In June 1979, 33 patches of cordgrass were recorded along the periphery of the sewage lagoons (Figure 15) an increase from the 24 patches recorded previously (see Clapper Rail Habitat: Requirements and Improvement project). By July 1980, a total of 103 patches were recorded. Many of the new patches consisted of fewer than 10 stems, and in three cases, only a single shoot of cordgrass, which were isolated by several meters from existing patches. This indicated that at least some of the patches had established from seed rather than vegetative reproduction. This was the first documentation of natural seedling propagation in several years of study of San Diego County salt marshes.

Associated research had indicated that the highest establishment from seeds for cordgrass was obtained by submerging seeds in water with low salinity. Interstitial soil salinity decreased dramatically in 1980 due to freshwater input from winter storms. Measurements in the marsh associated with Oneonta Slough documented this decrease and its effect on cordgrass density and height. The 101 0.25-m² quadrats sampled in the previous study, Clapper Rail Habitat: Requirements and Improvement, were re-sampled. In October 1979, overall mean soil

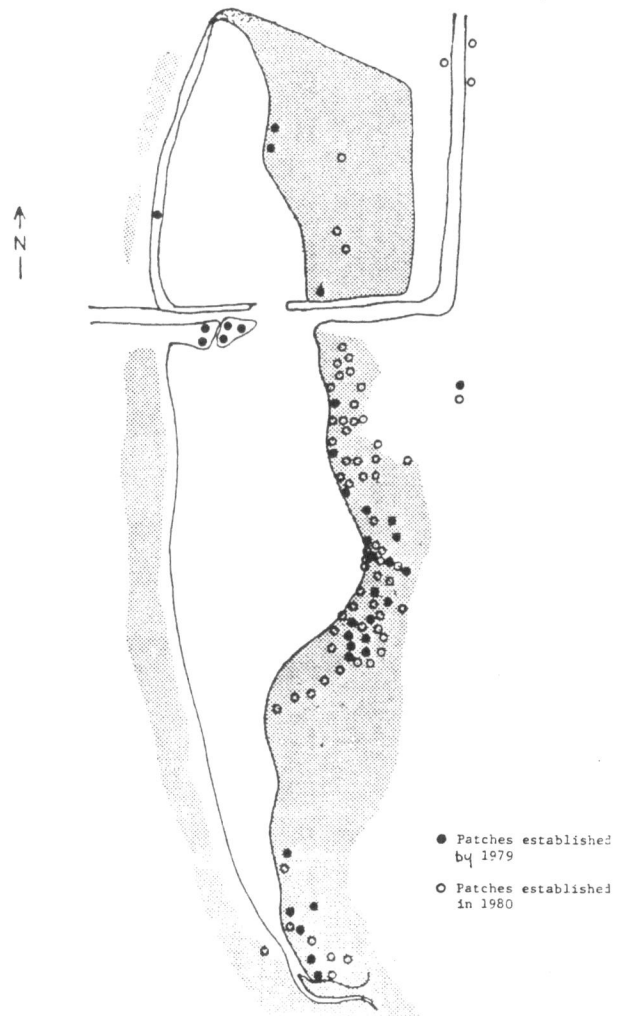

● Patches established by 1979

○ Patches established in 1980

FIGURE 15.
CORDGRASS PATCHES AT FORMER SEWAGE LAGOONS 1979–1980

salinity was 43.8 ppt. In April 1980, following February flooding, mean soil salinity had dropped to 14.8 ppt. By September 1980 it had risen again to 41.0 ppt. When height and density of cordgrass were compared for September 1980 and September 1979, there was an increase in density of 17 shoots/m², a 19-cm increase in mean height, a 30-cm

increase in maximum height and a 40% increase in biomass. These data suggested that a short term decrease in soil salinity had a positive effect on cordgrass growth.

LESSONS LEARNED

Although relatively simple in design, this study added significantly to the understanding of cordgrass response to a number of factors. Perhaps most significant was the successful establishment of cordgrass in areas that appeared to be suitable but did not support this species. This has had direct application to wetland restoration in Southern California where newly excavated marsh plains are planted with cordgrass and other species, depending on elevation. Prior to these experiments, this commonly accepted method of marsh establishment was unproven.

The response of cordgrass germination and growth to short term decrease in soil salinity is confounded by the fact that nutrient-rich soils are also conveyed by fresh water floods. While germination in the laboratory was most successful at low salinities, the response of cordgrass to flooding in the natural marsh may be a combined response to reduced salinity and nutrient enrichment.

Regardless, wet winters obviously have a positive effect on cordgrass growth and expansion. Decreases in soil salinity should be short term in order to produce these positive effects. During the same period (1980) long-term decreases in soil salinity at the San Diego River Estuary from major flooding and reservoir drawdown resulted in salt marsh species being replaced by cattails (*Typha* sp.). The salt marsh recovered after the freshwater input ceased.

The ability to study numerous, small patches of cordgrass has lent insight into the growth dynamics of the species with direct application to restoration. Small patches rapidly expand vegetatively to form clones that can be meters in diameter. Cordgrass plugs planted on two-meter centers typically coalesce into mature, continuous canopy in about three years.

Additional experiments on the addition of nitrogen to natural marshes were conducted by Dr. Zedler and her collaborators— Jordan Covin in Tijuana Estuary and Kathy Boyer in San Diego Bay. Cordgrass typically responded to addition of nitrogen in various forms but also demonstrated complex interactions with other species in nitrogen uptake and release and subsequent effects.

3 COASTAL WETLANDS MANAGEMENT: RESTORATION AND ENHANCEMENT

Principal Investigator:

Joy B. Zedler, San Diego State University, Department of Biology

Research Assistants:

Phil Williams, San Diego State University, Department of Biology

John Boland, San Diego State University, Department of Biology

Chris Nordby, San Diego State University, Department of Biology

Date: 1980

Funding Agency:

University of California Sea Grant College Program

Reported in:

Zedler, J.B., P. Williams, J. Boland and C. Nordby. 1980 Progress Report. NOAA Sea Grant Project. Coastal Wetlands Management: Restoration and Enhancement.

This study was conducted in parallel with the 1980 restoration work funded by the U.S. Navy and had similar goals:

1) determining what regulates the natural growth of salt marsh plant species; and

2) testing methods of propagating native salt marsh plants using existing populations of marsh plants as a source of propagules.

Regulation of Natural Growth of Salt Marsh Plant Species. The study was based on long-term observations of salt marsh dynamics at three regional wetlands which led to a conceptual model of how the interaction of fresh water and tides controls the structure and productivity of salt marsh vegetation in southern California. The model included three case studies.

Case 1. Fresh water input with no long-term alteration of tidal influence. Flooding in 1980 resulted in an unusually large input of freshwater, sediment and debris into the salt marsh of the estuary. Water remained in the estuary long enough to leach salts from the normally hypersaline soils. This decrease in salinity in combination with sediment and nutrient input was apparently responsible for the dramatic increase in the growth of cordgrass. Researchers recorded a 19-cm increase in average height, a 30-cm increase in maximum height and an average increase in density of 17 individuals/m². Biomass increased 40% compared to previous years. Fresh water also was credited with expansion of cordgrass distribution with seedling establishment observed for the first time.

Case 2. Fresh water input with blocked inlet. This case was based on observations of Los Peñasquitos Lagoon in 1978. While rainfall in 1978 did not lead to massive flooding as it did at Tijuana Estuary in 1980, inlet closure and freshwater impoundment leached salts from the soils much as it did in Tijuana Estuary in 1980. Productivity of Pacific pickleweed (*Salicornia virginica*) doubled in response compared to previous years.

Case 3. Continuous fresh water input in normally saline wetlands. This case was based on observations of the results of continuous reservoir drawdown on the salt marsh of the San Diego River. Prior to 1979, the marsh was a nearly monotype of Pacific pickleweed. Prolonged inundation with fresh water leached salt from soils, killed the monotypic pickleweed population, and provided conditions for invasion by cattails (*Typha* sp.). A gradual return to pickleweed was predicted once freshwater input ceased and tidal flows reestablished hypersaline conditions: however, with the advantage of hindsight, through transplantation of cordgrass plugs from Tijuana Estuary, the site is currently dominated by cordgrass and salt marsh daisy (*Jaumea carnosa*).

LESSONS LEARNED

Together these three case studies demonstrated how the structure and productivity of salt marshes in Mediterranean-type climates respond to fresh water input. Under typical conditions, salt marsh soils are hypersaline and vascular plant productivity is low (net above ground dry weight productivity of approximately $1km/m^2/year$). A small or temporary increase in freshwater can increase their productivity somewhat (Case 1); a prolonged influence increases their productivity greatly (Case 2); and a continual freshwater influence can change the entire character of the wetland from salt marsh to freshwater marsh (Case 3). The authors cautioned against making value judgments regarding changes in productivity as higher vascular plant productivity may result in a decrease in algal productivity within the same system. They further predicted that fresh water alters the entire ecosystem, affecting the type of available food sources and productivity of higher trophic levels.

Methods of Propagating Native Salt Marsh Plants. Because of the importance of cordgrass to wildlife such as the endangered light-footed Ridgway's rail, and because cordgrass rarely invades new habitats in southern California, efforts were focused on propagating this species from seed, rhizomes, and whole plants under a variety of environmental conditions. Seeds were stored at 5° C in fresh, brackish, and sea water. Average germination rate was 47%. Of a typical 100 seeds germinated in petri dishes 38 survived to be potted, 18 survived potting and 17 survived for field transplantation demonstrating that germination and early seedling survival were critical in rearing individuals to maturity. Salinity did not affect germination rates. Storage time did affect germination with highest rates recorded for seeds stored 0.5 months and lowest for seeds stored 7.5 months. Propagation from rhizome segments was unsuccessful and transplantation of whole plants from the field was identified as the most reliable technique for establishing cordgrass in the field (Zedler et al. 1980).

Propagation of cordgrass from seed is limited by the early stages of germination and seedling survival. Seeds should be stored at 5° C in water. The salinity of the water in which the seeds are stored (fresh, brackish, or sea water) does not appear to influence germination but does affect seedling survival with greater survival in fresh water. Optimal storage time for seeds was approximately two weeks with seeds losing viability with time stored.

Studies confirmed the results of the sister study that transplantation of whole plants collected from the field is the preferred method for establishing cordgrass on mudflats of the proper elevation for this species.

4 DUNE RESTORATION AT TIJUANA ESTUARY

Project Manager:
 Paul Jorgensen CSP
Principal Investigator:
 Joy B. Zedler, San Diego State University, Department of Biology,
 Pacific Estuarine Research Laboratory
Research Assistant:
 Brian Fink, San Diego State University, Department of Biology,
 Pacific Estuarine Research Laboratory
Date: 1985
Reported in:
 Jorgensen, P. Tijuana River National Estuarine Sanctuary, Dune Restoration at
 Tijuana Estuary, April 1987.

This project was initiated to help restore the barrier dune system at Tijuana Estuary following dune overwash and inlet closure in 1984. Following that event, the USFWS dredged Oneonta Slough using a clam shell dredge and rebuilt approximately 8.5 acres of dunes along 3,000 feet of barrier beach in the northern arm of the estuary. Approximately 60,000 yd³ of sand was moved resulting in a recreated dune system approximately 10 feet above mean sea level (MSL). The dune system was artificially shaped to form a continuous berm approximately 200 feet wide at the base

and 10 feet high to provide protection from future overwash.

The south arm of the estuary was also dredged. Approximately 1,500 feet of channel was dredged and over one mile of dunes was created, totaling about 60,000 yd³ of sand and covering approximately 25 acres. About 900 feet of channel remained to be dredged by the time the report was submitted.

Sand deposited in the southern arm was pushed towards the beach to form staggered rows of mounds, each about 30 feet at the base and six feet above grade, which varied from nine–13 feet above MSL. The mounded design was based on natural dunes at San Quintín, Laguna Figueroa and Estero Punta Banda in Baja, California.

METHODS

In the north arm, four native species selected for their hardiness and ability to build and stabilize dunes were planted from nursery-grown stock, including red sand verbena (*Abronia maritima*), beach-bur (*Ambrosia chamissonis*), coastal goldenbush (*Haplopappus venetus = Isocoma meziesii*) beach evening-primrose (*Camissionia cheranthifolia*). An irrigation system consisting of two parallel lines with drip emitters every 10 feet was installed prior to planting. In all, about 1,400 potted plants and seed were installed. By species, 330 plants and seed of red sand verbena, 785 pots of beach-bur, 250 pots of coastal goldenbush and seed of beach evening-primrose were installed. Various pot sizes were utilized, mostly from two-inch liners and one-gallon pots. Plants were grown at PERL.

Plants installed in the south arm were similar to those in the north with the exception that coastal goldenbush was omitted and beach sand verbena (*Abronia umbellata*) was added. Plants were grown in 14-inch deep "super cells" and one-gallon pots at PERL. About 50–75 gallons of red-sand-verbena seed with fruit were collected locally and distributed along the entire dune. Plantings consisted of beach-bur (1,500 pots and seed), red sand verbena (1,000 pots and seed), beach sand verbena (100 pots) and beach evening-primrose (900 pots).

Approximately 50% of the project was set up as an experiment to test five sets of variables. These included:

1) vegetated, rebuilt dunes with fabric fencing;
2) vegetated, rebuilt dunes without fabric fencing;
3) rebuilt dunes without fabric fencing and without revegetation;
4) revegetation with fabric fencing and no rebuilt dunes; and
5) no treatment (control).

Study plots were 200-foot-long sections of dunes with replicates for each experimental treatment. Planting in experimental plots consisted of 20 plants of each species. Therefore, planting densities were low.

RESULTS

At the time the report was submitted, restoration work was not completed. However, some results were available from work done in the north arm. In the north arm, the berm–dune was partially washed away on the seaward side and ocean washovers occurred along approximately two-thirds of the constructed dune. This accounted for the loss of approximately 70% of the potted plantings and many germinated seedlings.

Although planting was conducted in mid-winter there were dry, heavy winds the day following two of the three major planting efforts. This led to high mortality although shifting sands and overwash made it impossible to mark individual plants and quantify mortality.

Qualitative observations showed that virtually all coastal goldenbush died. Except for the physical effects of overwash, both red sand verbena and beach-bur had survival rates of 50% or greater. All seeded areas appeared to germinate well, especially red sand verbena.

At the time of the report, it was estimated that the northern berm–dune had roughly 25% vegetative cover with 90% of that 25% comprised of non-native, naturally recruited European sea-rocket (*Cakile maritima*). Red sand verbena and beach-bur contributed to the remaining 10%.

Few results were available from the southern dune project. However, mortality estimated at 80% occurred just after planting due to high winds.

LESSONS LEARNED

Many lessons were learned from the dune restoration project. It was determined that the northern berm–dune was constructed too far west into the surf zone. Waves destroyed the irrigation lines and precluded fencing the west side of the dunes. In other areas, the irrigation was vandalized and needed constant repair.

It was determined that seed application was the best method of establishing plants. Seed is cheaper to collect and install and can germinate when conditions are optimal as opposed to plantings which are subject to uncontrollable environmental factors. Red sand-verbena seed germinated 16 months after it was scattered in the northern berm–dune.

Irrigation is advisable but budgets should account for the manpower needed to maintain the lines and emitters. The non-native European sea rocket will invade aggressively in any dune restoration project. Fencing is required to prevent trampling. Again, project budgets should account for maintenance from potential vandalism.

5 DUNE VEGETATION REESTABLISHMENT AT TIJUANA ESTUARY

Principal Investigator:
> Joy B. Zedler, San Diego State University, Department of Biology, Pacific Estuarine Research Laboratory

Research Assistant:
> Lisa F. Woods, San Diego State University, Department of Biology, Pacific Estuarine Research Laboratory

Date: 1987

Funding Agency:
> National Oceanic and Atmospheric Administration; TRNERR

Reported in:
> Wood, L.F. and J.B. Zedler. 1987. Dune vegetation at Tijuana Estuary: interactions between the exotic annual, *Cakile maritima* and the native perennial, *Abronia maritima*. NOAA Technical Memorandum. NOS MEMD.

This project and associated dune restoration projects by Brian Fink at Tijuana Estuary were undertaken in response to El Niño events that resulted in dune overwash and sedimentation of channels in 1983. In January 1983, high tide and a coincidental El Niño storm event washed dune sand into the channels of the estuary eventually resulting in closure of the inlet, as described earlier. As a result of the overwash, an extensive dune restoration was initiated on the coastline south of the inlet. The purpose of the project was two-fold: to return sand to its pre-overwash location and to stabilize the dunes system and to conduct research on dune plant interactions.

METHODS

In fall 1986, over 3,000 cubic meters (m³) of sand were bulldozed from the landward side of the coastal strand to the seaward side. The sand was dumped and left as hummocks approximately four meters wide at the base and two meters in height. In winter 1987, approximately 2,000 plants were planted on the hummocks. Planted species included species proven to provide stabilization, such as red sand verbena (*Abronia maritima*) as well as species with aesthetic or cultural value, such as dune primrose (*Oenothera cheiranthifolia*). Several hundred meters of plastic sand fencing were installed in an effort to trap sand and prevent trampling until the plants could establish. Unfortunately, in addition to the planted native species, the hummocks developed a robust population of invasive European sea-rocket (*Cakile maritima*). The purpose of this study was to investigate the role of sea-rocket on the restored dunes. The dune plant community at Tijuana Estuary was compared with that of two less disturbed dunes at San Quintín, Baja California, Mexico. Percent cover by sea-rocket on the restored dunes was compared to that of areas at Tijuana Estuary left undisturbed. The effect of sea-rocket on growth and establishment of the former dominant, dune primrose, and the nature of the interaction was investigated.

RESULTS

Sea-rocket was the most common species at the restored and undisturbed dunes at Tijuana Estuary. By comparison, dune primrose dominated the populations at San Quintín. It was concluded that the presence of dune primrose had little effect on the distribution of sea-rocket. In an artificial dune experiment conducted at PERL, sea-rocket showed a significant response to induced competition with dune primrose; however, the nature of that interaction was not discerned. A fertilizer experiment designed to determine the mechanism of the interaction was inconclusive. Dune primrose is likely more limited by disturbance than by the presence of sea-rocket.

LESSONS LEARNED

Sea-rocket is an aggressive exotic annual plant that does very well in disturbed areas. Anthropogenic disturbances, such as trampling and the introduction of non-native species, are factors in the decline of red sand verbena at Tijuana Estuary.

Because anthropogenic disturbances promote sea-rocket and discourage the perennial red sand verbena, the most stable dunes can develop where anthropogenic disturbances are kept to a minimum. In areas already dominated by sea-rocket, it is even more important to minimize disturbances.

Because sea-rocket is a poor sand stabilizer, physical disturbances have a greater effect in areas which are dominated by it than in areas which are protected by red sand-verbena. Sea-rocket seedlings preferentially germinated in disturbed areas whereas sand verbena germination is inhibited by high disturbance levels. Thus, sea-rocket aggressively invades disturbed areas, then prevents the subsequent establishment of red sand verbena through direct inhibition and by failing to buffer disturbances to the substrate.

In addition to its effect on strand stability and on native plants, sea-rocket may also have an adverse impact on the insect community. It has been suggested that bare areas on the dune are an essential resource to several substrate-dwelling species. At Tijuana Estuary, the percent cover of sea-rocket on the bulldozer-created hummocks was 75%, whereas at sites dominated by red sand-verbena the maximum cover recorded was much lower. In areas with a high percent cover, sea-rocket may have a negative impact on dune insects requiring open space.

6 BORDER FIELD STATE PARK DUNE RESTORATION PROJECT

Project Manager:
> Paul Jorgensen CSP

Principal Investigator:
> Joy B. Zedler, San Diego State University, Department of Biology,
> Pacific Estuarine Research Laboratory

Research Assistant:
> Brian Fink, San Diego State University, Department of Biology,
> Pacific Estuarine Research Laboratory

Date: 1989

Reported in:
> Border Field State Park Dune Restoration Project, Final Report, June 30, 1991,
> by Brian Fink, Pacific Estuarine Research Laboratory, for California Department
> of Parks and Recreation.

This project represented another attempt to restore dune habitat, as well as coastal strand and coastal bluff habitats in the south end of the estuary. Nineteen plant species were included in the plantings, 12 of which were not formerly present. The project planted red sand verbena, beach sand verbena, beach-bur, shadscale (*Atriplex canescens*), big saltbush (*Atriplex lentiformis*), tread lightly (*Cardionema ramosissima*), coastal goldenbush, coastal sagebrush (*Artemisia californica*), seashore morning-glory (*Calystegia soldanella*), false goldenstar (*Chrysopsis villosa* var, *sessiflora* = *Heterotheca sessiliflora* ssp. *sessiliflora*) saltgrass (*Distichlis spicata*), Orcutt's dudlyea (*Dudlyea attenuata*), California encelia (*Encelia californica*), bluff buckwheat (*Eriogonum parvifollium*), cliff spurge (*Euphorbia misera*), Palmer's frankenia (*Frankenia palmeri*), San Diego marsh-elder (*Iva hayesiana*), spiny rush (*Juncs acutus*), prostrate lotus (*Lotus nuttallianus* = *Acriispon prostratus*), and California desert thorn (*Lycium californicum*).

One of the intents of the project was to increase the diversity of plants which grow in coastal habitats at Border Field State Park, including coastal strand, backdune, protected coastal bluff and unprotected coastal bluff. As the plant species that may have once inhabited these habitats at BFSP were never formally documented, this list was derived from plant species present at Silver Strand State Beach and the seaside bluffs at Torrey Pines State Reserve in San Diego, California.

METHODS

The 1991 Final Report references a 1990 interim report that is not available regarding methods. However, it is evident that selected species were planted in a variety of habitats with varying degrees of exposure to winds and other maritime stressors, such as salt spray and dune overwash. Sand fences were erected in experimental plots to provide protection from these stressors and to stabilize dune sand. Saltgrass was planted in patches as a nurse plant to provide further protection from stressors and to reduce plant herbivory by rodents. Fertilization and seeding experiments were conducted; however, the 1991 report offers no details as to the methods employed. Beach wrack was employed to buffer plants from harsh winds and salt spray.

Seven belt transects of unreported length were used to estimate percent cover and frequency of occurrence of naturally occurring dune vegetation in the project area. These transects established outside of experimental plots to characterize the condition of the site.

RESULTS

Survivorship of woody scrub species was generally higher in more protected microhabitats, including protected bluff, strandline protected by sand fences and within saltgrass patches. For example, coastal goldenbush showed 88% survival within protected sand plain and 89% within wind protected bluff compared to 25% survival in unprotected sand plain and 23% in unprotected bluff. Bluff buckwheat had 80% survival within unprotected sand plain; 57% within protected sand plain; 22% within unprotected bluff and 89% within protected bluff.

Beach wrack proved effective in stabilizing dune sand in areas that were not fenced, although no details were provided. Beach wrack also curtailed the growth of weeds in experimental plots.

Fertilization and seeding experiments were inconclusive as all planted species were destroyed. Estimates of cover revealed that the non-native European rocket was the dominant species in all transects.

LESSONS LEARNED

As with other dune restoration projects at Tijuana Estuary, fencing to preclude trampling was recommended. Sand overwash areas should be planted with saltgrass and beach morning-glory. Saltgrass should be heavily fertilized; beach morning-glory should not be fertilized. After saltgrass has formed small hummocks, they should be seeded with red sand verbena and beach-bur.

Exposed bluff areas should be planted with maritime succulent scrub species, such as cactus (*Opuntia* spp.), beach buckwheat, Palmer's frankenia and California desert thorn. Protected backdune areas that are removed from storm surge are conducive to establishment from seed, especially

sand verbena species, prostrate lotus, and tread lightly.

Planting one-gallon container stock was recommended with frequent watering during the first month followed by infrequent watering (once per month). Fertilization is recommended for most species as sand contains little or no nutrients.

7 TIJUANA ESTUARY TIDAL RESTORATION PROGRAM (TETRP)

Project Author:
> Joy B. Zedler, San Diego State University, Department of Biology, Pacific Estuarine Research Laboratory

Technical Coordinator:
> Ted P. Winfield, ENTRIX, Inc.

Technical Assistants:
> Ted Griswold, San Diego State University, Department of Biology, Pacific Estuarine Research Laboratory
> Sharon Lockhart, San Diego State University, Department of Biology, Pacific Estuarine Research Laboratory

Project Engineers:
> Berryman & Hennigar

Project Contractor:
> Roberts Engineering

Project Administrator:
> Mayda Winter, SWIA

Construction Manager:
> Chris Nordby, Tierra Environmental Services

Date: 1991–2004

Funding Agency:
> California State Coastal Conservancy

Reported in:
> ENTRIX, Inc., Pacific Estuarine Research Laboratory and Phillip Williams and Associates. 1991. Tijuana Estuary Tidal Restoration Program Volume I: Draft Environmental Impact Report/Environmental Impact Statement.

The Tijuana Estuary Tidal Restoration Program (TETRP) was designed in response to the sedimentation issues affecting primarily the southern arm of the estuary but also similar issues in the northern arm. The project was conceived shortly after the tidal inlet closed in 1984 when it was realized that the estuary was vulnerable to cataclysmic desiccation due to lack of tidal influence. Thus, the purpose of the project was to define and implement a restoration program that would assure the long-term protection of the valuable estuarine ecosystem. Conceptual engineering plans and detailed analysis of existing conditions and potential environmental impacts from project implementation were presented in a project EIR/EIS completed in October 1991.

A key element of TETRP was to return the estuary to an historical condition with self-sustaining tidal flushing. The EIR/EIS documented the decline in tidal prism of the estuary from an estimated 1,550 acre-feet in 1852 to 290 acre-feet in 1989. In addition, the tidally-influenced portion of the estuary declined from approximately 870 acres in 1852 to 330 acres in 1989. Much of these losses occurred in the southern arm of the estuary where channels were constricted by sediment and the marsh plain was approximately two feet higher in elevation than the marsh plain in the north arm. In the northern arm, dune overwash during coincidental storm events and higher than usual tides associated with El Niño events resulted in deposition of sand in the main Oneonta Slough tidal channel further compounding loss of tidal prism.

TETRP was designed to include two primary elements: The Model Project and a 495-acre Restoration Project. The Model Project, to be implemented first, was composed of three components:

1) Oneonta Slough Widening;
2) construction of the Connector Channel; and
3) construction of the 20-acre experimental marsh (Model Marsh; Figures 16 and 17).

Oneonta Slough Widening. Widening of Oneonta Slough was proposed to keep the north arm of Tijuana Estuary open as the barrier beach migrated inland over time. A hardpan was identified along the east bank of the main slough channel near the tidal inlet that restricted eastward migration. The EIR/EIS proposed removal of approximately 26,000 yd^3 of hardpan to alleviate this condition.

Connector Channel. The Connector Channel (also known as the Tidal Linkage) was designed to improve tidal circulation by connecting the northern part of Oneonta Slough with the former sewage lagoons, renamed as the tidal lagoons, located in the northeastern part of the north arm. The Connector Channel would accomplish several purposes, including:

1) stabilizing the channel system in the northern arm and reducing sedimentation in the tidal channels;
2) confining human access in the northern arm;

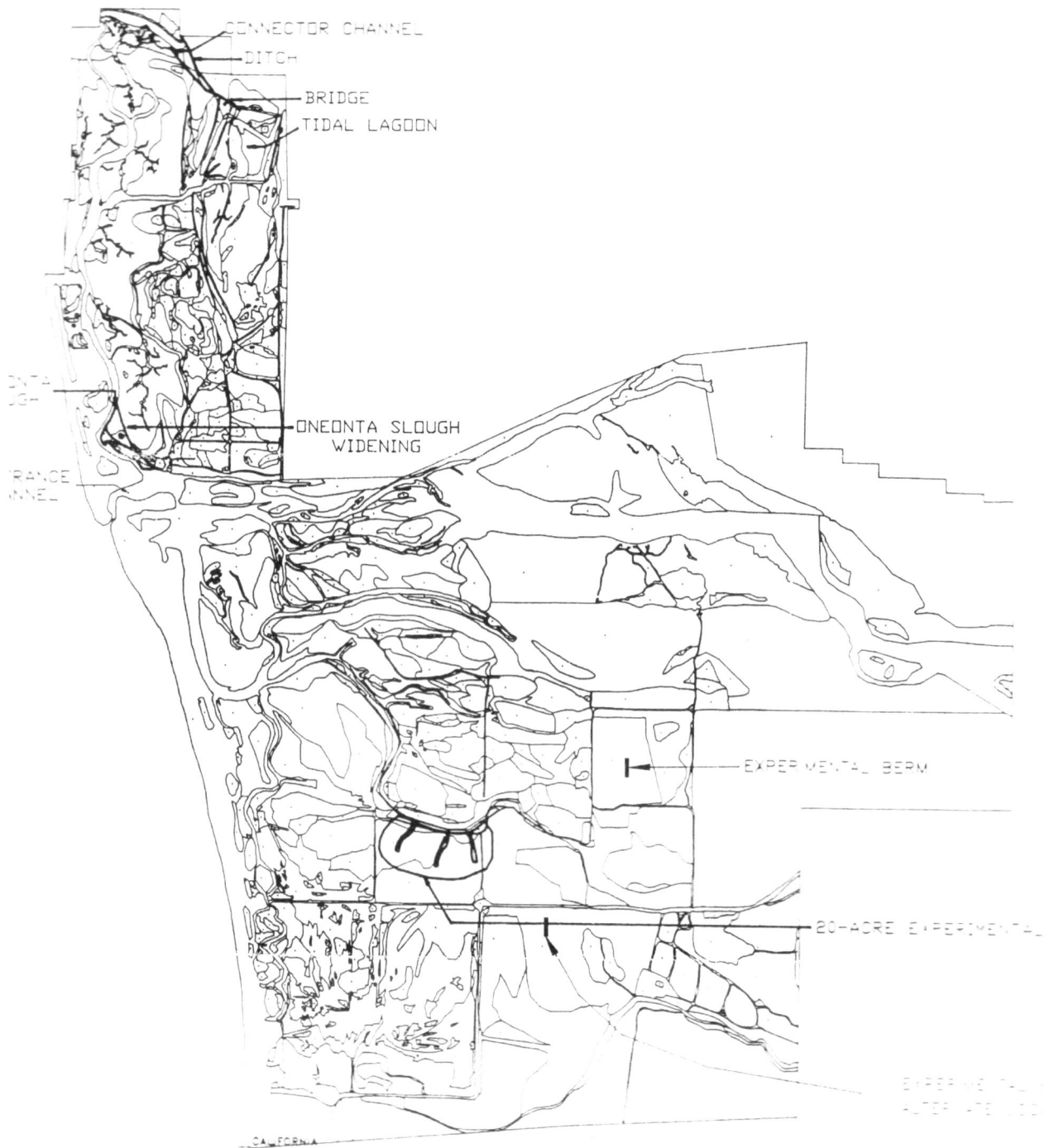

Labels on figure: CONNECTOR CHANNEL, DITCH, BRIDGE, TIDAL LAGOON, ONEONTA SLOUGH WIDENING, EXPERIMENTAL BERM, 20-ACRE EXPERIMENTAL, CALIFORNIA

FIGURE 16.
CONNECTOR CHANNEL, ONEONTA SLOUGH WIDENING
AND MODEL MARSH FROM 1991 EIR/EIS

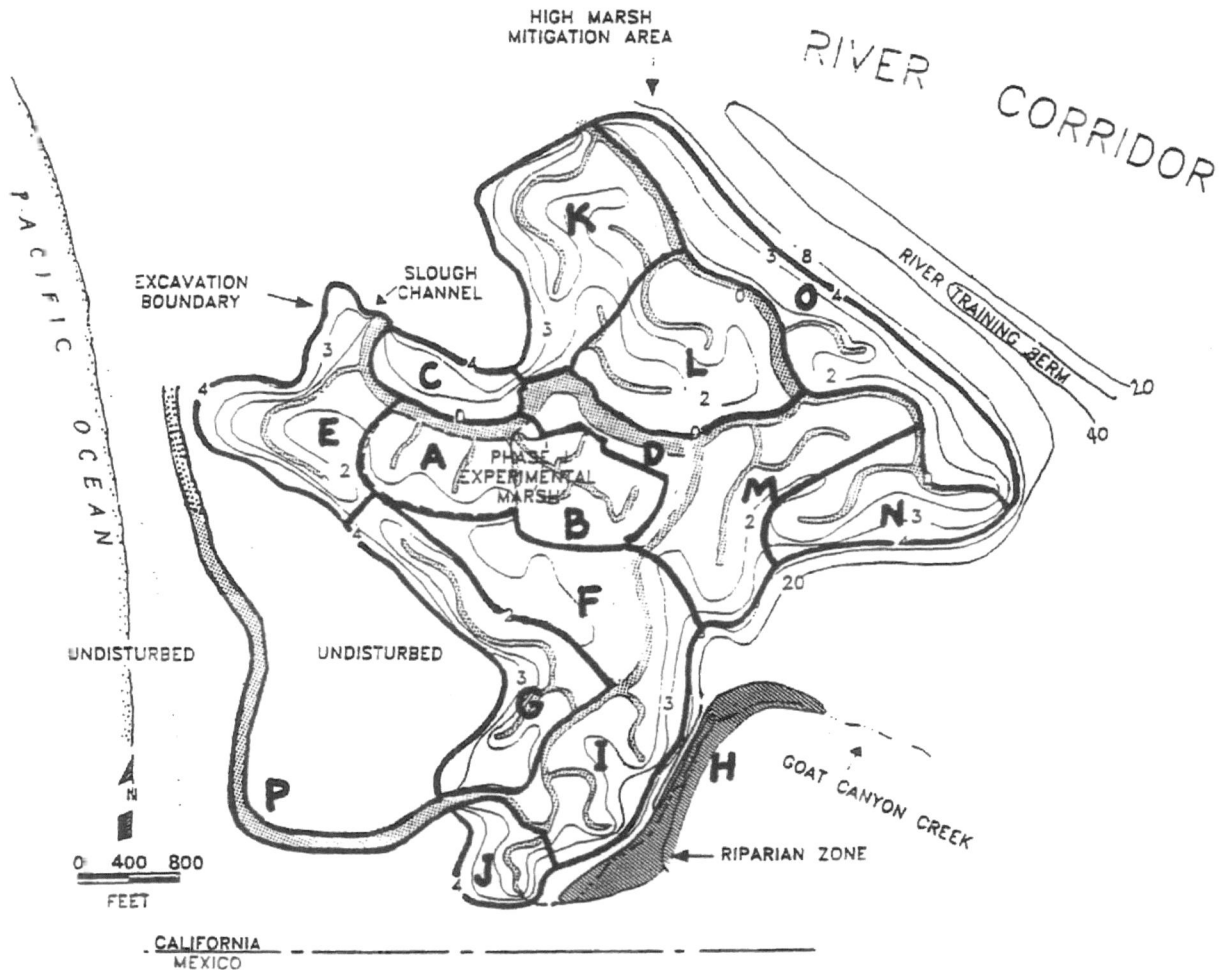

FIGURE 17. MODEL MARSH AND 495-ACRE RESTORATION FROM 1991 EIR/EIS

3) providing additional salt marsh;
4) collecting low-flow street runoff near the visitor center prior to reaching the salt marsh; and
5) providing an interpretive opportunity immediately adjacent to the visitor center. A 30-foot long timber bridge would be constructed over the connector channel to provide access to visitors.

20-Acre Model Marsh. The 20-acre model marsh (sometimes called Friendship Marsh which included the surrounding planned restoration area) consisted of two components:

1) construction and monitoring of an experimental tidal marsh; and
2) construction and monitoring of an experimental berm from the sediment excavated to construct the marsh.

The 20-acre experimental marsh was designed to address the hypothesis that ecosystem development could be accelerated by increasing topographic heterogeneity. The experimental berm would address the suitability of existing soils to construct river training structures required for the planned 495-acre restoration. Four disposal alternatives were considered:

1) onsite disposal in the river training structure;
2) ocean dumping;
3) beach disposal; and
4) other offsite disposal locations.

495-Acre Restoration Project. The 495-acre project consisted of four components:

1) restoration of 495 acres of tidal marsh in the south arm;
2) construction of a river training structure;
3) stabilization of barrier dunes; and
4) restoration of riparian habitat.

Restoration of the 495-acre marsh would be done in phases. The river training structure would protect the restored marsh from sedimentation from the Tijuana River. The total footprint of the 495-acre marsh and river training structure would be between 507 and 540 acres depending on the alternative river training berm selected.

Tidal Marsh Restoration. Restoration of the tidal marsh would involve excavating a series of tidal channels, lowering the existing marsh plain, and construction of either a river training berm or levee. The volume of material removed for the total tidal marsh restoration was estimated to be 6,000,000 yd^3. The mean diurnal tidal prism of the estuary after implementation would be approximately 845 acre-feet, or about one half the tidal prism estimated from the 1852 map of the estuary.

River Training Structure. The river training structure was deemed necessary to prevent the Tijuana River from migrating into the restored marsh and depositing sediment. The river training structure would extend from Spooner's Mesa into the estuary in a west–northwest direction.

There were two alternative river training structures: an erodible berm or a rip-rapped levee. The erodible berm would be 5,000 feet long, 25 feet high, and 400 feet wide at the base. The berm footprint would be 46 acres and would withstand 100 years of average erosion rates. Slopes would be planted with coastal sage scrub plant species.

The rip-rapped levee would be 5,000 feet long, 10 feet high and 100 feet wide at the base. The eastern slope would be protected with rock rip-rap. The levee footprint would be 12 acres.

Sand Dune Stabilization. The EIR/EIS acknowledged the need for a dune stabilization plan but deferred this to a later date.

Riparian Restoration. A minimum of 100 acres of willow-dominated riparian habitat restoration was identified in the

FIGURE 18. CONNECTOR CHANNEL

EIR/EIS. This was planned in an area that, at the time of the EIR/EIS, was devoted to sod farming. That area is currently (2016) fallow agricultural fields. An additional 15 acres of riparian restoration was planned for Goat Canyon which now supports the sedimentation basins.

Components of TETRP Implemented. Two of the three components of the Model Project were constructed although with modifications to the original design—the Connector Channel (Tidal Linkage) and the 20-acre Model Marsh. The Tidal Linkage Project was funded by the State Coastal Conservancy and the USFWS. The Model Marsh was funded by the State Coastal Conservancy. Both incorporated extensive research components in their design that tested hypotheses on coastal wetland restoration as well as lessons learned during project permitting, construction and long-term monitoring.

Tidal Linkage (Connector Channel) Reported in: Year Three Monitoring Report, Oneonta Slough Tidal Linkage Project, Tijuana Estuary, California. Tierra Environmental Services, October 6, 1999.

Although referred to as the Connector Channel in the EIR/EIS and construction bid documents, this component was most often referred to post-construction as the Tidal Linkage and that name is retained here. The Tidal Linkage is located in the northwestern

end of Tijuana Estuary immediately adjacent to the Visitors Center (Figure 18).

METHODS AND RESULTS

The 1.84-acre restoration was constructed from approximately mid-December 1996 through April 1997. The project connected then terminal tidal creeks in northwestern Oneonta Slough with the former sewage lagoons (renamed "Tidal Lagoons") located east of Oneonta Slough, hence the original name of Connector Channel. The project was designed to salvage all plants and soils that might be useful in restoration of a wetland. The majority of Tidal Linkage was located on upland habitat with each end connecting to existing wetlands. Thus, impacts to existing wetlands were limited to the excavation of narrow channels. A total of 0.07 acre of salt marsh habitat was permanently impacted by the project and 0.17 acre of tidal channels was temporarily impacted. Upland soils were deemed compatible with beach nourishment and were discharged as a slurry via a long pipeline to the surf zone. Wetland soils excavated as part of the project were salvaged and spread on the restored northern tidal plain to facilitate plant establishment and growth, while material excavated from the sewage lagoon was imported to the south, experimental tidal plain.

In response to researcher request, the Tidal Linkage included a sinuous main channel with a salt marsh plain on each side that included low, mid- and high salt marsh elevations (Figure 19). Plants salvaged from existing salt marsh at the west and east end of the channel consisted of eight-inch square blocks, or "plugs," of cordgrass and six-inch square blocks of mixed mid-elevation marsh species. In addition, plugs of saltgrass (*Distichlis spicata*) and individual Parish's pickleweed (*Arthrocnemum subterminale*) were salvaged and planted. In all, more than 1,300 cordgrass blocks were planted on three-foot centers, 2,730 blocks of mixed marsh species were planted on 2.5-foot centers, 1,936 plugs of saltgrass and eight individual pickleweed plants were planted. Approximately 2,000 ft^2 of salt marsh and 80 yd^3 of marsh soils were salvaged. The south tidal plain was planted with eight native salt marsh species in 87 two-meter x two-meter plots with zero–six species per plot.

The project was constructed from west to east with the existing tidal creek extended by a small Bobcat excavator (Figure 20). The excavator was fitted with a customized open-bottomed bucket that allowed salvage of approximately two-foot x two-foot x one-foot deep blocks of mid-elevation salt marsh plants which were later divided in six-inch square blocks.

Cordgrass that occurred in the path of the excavated channel was harvested by hand as plugs and potted in one-gallon plant containers. High salt marsh species, e.g., saltgrass, in the channel's path were salvaged by front loader as sod blocks. All plants were stored on-site in plastic lined trenches and watered with fresh water until excavation was completed.

Once the tidal creek was extended to the

FIGURE 19.
CONNECTOR CHANNEL 1996

FIGURE 21.
CONTRACTOR IN SEWAGE LAGOON

FIGURE 20.
BOBCAT EXCAVATOR WITH
CUSTOMIZED BUCKET

Tidal Linkage marsh plain, further excavation was conducted with conventional, full-size excavators. The material excavated was placed in a deep, water-filled pit where a small cutterhead dredge created a slurry that was pumped approximately one mile to the beach for disposal in the upper surf zone. Approximately 18,000 yd^3 of excavated soils were disposed of in this manner.

Regular bacteriological monitoring of the receiving waters was a requirement of the USACE Section 404 permit for the project. The eastern end of the Tidal Linkage extended well into the northern Tidal Lagoon, the substrate of which consisted of very soft sediments from their former use as sewage oxidation ponds. The contractor constructed an earthen berm to the eastern terminus and excavated and hauled out both the soft channel and berm material (Figure 21).

Once the tidal plain and channel were excavated, the salvaged salt marsh soils were spread on the exposed soils of the

north tidal plain and salvaged salt marsh plants were planted in holes augered into the surface. Low, mid- and high salt marsh species were planted on the entire north side of the channel and on a portion of the south side. Most of the south tidal plain was reserved for experimental plantings by researchers from PERL. The area set aside for research included 0.02 acre for low marsh, 0.25 acre for mid- marsh, and 0.04 acre for high marsh. An irrigation system with rotary sprinkler heads was installed along both the north and south high marsh area to facilitate plant establishment.

The south marsh plain of the Tidal Linkage was planted as an experiment to distinguish between species that needed to be planted from those that colonized naturally and to determine if ecosystem functions could be accelerated by planting species-rich assemblages. This is presented in greater detail under Experiments to Restore Diversity.

The northern tidal plain was monitored quarterly for the first two years and semi-annually for the third year. At the end of the third year it met the compliance criteria, the permitting agencies signed-off on the project and monitoring was discontinued.

EXPERIMENTS TO RESTORE DIVERSITY

As stated previously, the south marsh plain of the Tidal Linkage was planted as an experiment to determine if ecosystem functions could be accelerated by planting

FIGURE 22. EXPERIMENTAL PLOTS AT CONNECTOR CHANNEL

species-rich assemblages. The experiment had five blocks with 15 two-meter x two-meter plots/block (Figure 22). The plots were planted with zero, one, three or six native halophytes randomly selected from a pool of eight native species. Planted treatments had 90 seedlings spaced on a 20-centimeter grid. Due to random selection of each species, all eight species were planted at approximately equal quantities (790 seedlings/species). Additional test plots were established to assess the role of processed kelp as a soil amendment on plant establishment and growth.

LESSONS LEARNED

There were numerous lessons learned from monitoring the restoration of the Tidal Linkage project that can be applied to future restoration efforts. Some could be anticipated from earlier restoration work while others were unexpected. One of the possibly anticipated lessons was responding to herbivory by rabbits. Cordgrass was grazed by rabbits immediately after planting. Installation of two-foot high aviary wire fences along the upland margins on both sides of the channel effectively controlled the problem. However, herbivory by American coots (*Fulica americana*) was not anticipated. Coots selectively grazed pickleweed planted from salvaged salt marsh blocks. Because they entered the site from the channel and because they can fly, fencing was not expected to control them, but it seemed to reduce their numbers. As the marsh matured, grazing by coots decreased and pickleweed recovered.

Another unanticipated challenge involved smothering of marsh plants by algae. Shortly after construction was completed and tidal influence was established, large floating mats of *Enteromorpha* (=*Ulva* spp.) blanketed the planted marsh species and had to be removed by hand. Plants beneath the dense algal mats began to pale in color and shaded marsh plants. The algal mats proved to be temporary and by summer were no longer as prevalent.

Monitoring of coliform bacteria at the surf zone discharge site provided another unexpected lesson. High bacterial counts, exceeding allowable levels by several times, were encountered throughout the project, although data from testing could not be located for this document. The RWQCB determined that the effects were localized at the point of discharge and were rapidly diluted. Discharge was allowed to continue with signage installed at the discharge pipe warning beach users to avoid the area.

From a construction management perspective, the project provided other valuable lessons. The contractors incorporated novel construction methods that facilitated excavation and plant salvage. The use of the Bobcat excavator to construct the connection to the west with the terminal tidal creeks made use of portable ¾-inch plywood sheets to provide a stable surface over the unstable marsh mud. These worked for the most part; however, the excavator still slid off the wooden surface and became buried in the soft substrate. The customized bucket for salvaging mid-salt marsh worked well and expedited that portion of the project. The use of the small dredge to create the slurry for disposal in the surf zone was novel, and although that dredge was subsequently destroyed by fire, a similar approach was recently implemented by the San Diego Unified Port District in its 2011 restoration and enhancement of the Chula Vista Wildlife Reserve in south San Diego Bay.

The Tidal Linkage required several in-field decisions for the project team, including modifications to the slopes of the high marsh to upland transition, densities of plants and contractor change orders. It was determined that the 2:1 slopes for transitioning to upland called out in the construction drawings were collapsing due to the high percent of soft sand on-site. The decision to modify to 3:1 slopes required that an additional 900 yd³ of soil needed to be removed resulting in a small change order of approximately $10,700. Just prior to planting the marsh plain, the USFWS

informed the project team that light-footed Ridgway's rails had begun nesting in early February in other regional wetlands and they wanted work in the existing low marsh at both ends of the project to be terminated by February 17, 1997. This put considerable pressure on the contractor and resulted in the project team changing the density of cordgrass blocks from two-foot centers called for in the construction documents to three-foot centers to expedite this work task. As the project neared completion, the contractor claimed that the quantity of soil that was estimated in the construction drawings was inaccurate and that there was more soil than estimated and requested another change order. The project team, in consultation with SWIA's attorney, countered that the original estimate was accurate and the change order was denied. This situation would also occur during the Model Marsh restoration with a different outcome. The lesson learned from both suggests very careful pre-construction surveys of existing elevations from which accurate earthwork calculations can be made.

As monitoring of the site continued, it became evident that the sinuous channel that was designed to mimic more natural channels was eroding along its edges and the low marsh habitat was disappearing. The lesson learned from this was that restoration projects should be planned by a multi-disciplinary team of ecologists, civil engineers and hydrologists to design and model features like tidal creeks and channels.

Based on the results of the PERL experimental plantings on the south marsh plain coupled with long-term marsh-wide monitoring at Tijuana Estuary, pickleweed was excluded from plant pallets in subsequent restorations and two short-lived species (Bigelow's pickleweed and sea blite) proved able to self-recruit where there was a source of seed. Pickleweed has continued to establish naturally and, often, aggressively. It was also shown that planting six species led to complex canopies, greater biomass and nitrogen accumulation over the first few years, but that plots gradually converged on three-species assemblages, dominated by Pacific pickleweed, alkali heath and salt marsh daisy after a decade. The field experiment was replicated in a greenhouse resulting in a wealth of information on the interactions of the eight native marsh plain species. In general, planting more species did not result in more biomass, but planting a productive species, such as Pacific pickleweed, did. The use of soil amendments showed improved plant establishment and growth.

For more information on experiments associated with the Tidal Linkage please see:

Lindig-Cisneros, R. and J. B. Zedler. 2002. Halophyte recruitment in a salt marsh restoration site. Estuaries 25:1174-1183.

Keer, G., and J. B. Zedler. 2002. Salt marsh canopy architecture differs with the number and composition of species. Ecological Applications 12:456-473.

Callaway, J. C., G. Sullivan, and J. B Zedler. 2003. Species-rich plantings increase biomass and nitrogen accumulation in a wetland restoration experiment. Ecological Applications 13:1626- 1639.

Sullivan, G., J. Callaway and J. B. Zedler. 2007. Plant assemblage composition explains and predicts how biodiversity affects salt marsh functioning. Ecological Monographs 77:569-590.

Bonin, C. L., and J. B. Zedler. 2008. Plant traits and plasticity help explain abundance ranks in a California salt marsh. Estuaries and Coasts 31:682-693.

Model Marsh (Friendship Marsh) Reported in: Final Monitoring Report for the Friendship Marsh, Border Field State Park, California. Tierra Environmental Services. February 6, 2004.

The 20-acre Model Marsh was constructed during fall and winter of 1999–2000 (Figure 23). The location of the marsh was moved from the original design as the channel that would have provided tidal waters became choked with sediment. In addition, the experimental berm was not constructed. Sediment excavated from the Model Marsh was first de-watered on-site, then stockpiled in a disturbed area of Goat Canyon and ultimately used to restore an abandoned sand and gravel quarry (Fenton Quarry). An estimated 133,000 yd^3 was transported to the stockpile and stored during the 2000 bird breeding season (February 15– September 1). During the 2000–2001 non- breeding season, the sediment was used to fill the quarry which was subsequently planted with maritime sage succulent scrub plant species.

METHODS

The Model Marsh restoration was designed as a large field experiment with associated research (adaptive restoration). The 20-acre site included areas of mudflat, and low, mid- and high salt marsh. The Model Marsh was constructed as six experimental units with three replicate units with excavated tidal creek networks and three without (Figure 24). The design tested the hypothesis that salt marsh vegetation, as well as fish and invertebrate populations, would develop faster in areas with tidal creeks than without. Within the six experimental units, mudflats were intended for the lowest elevations and low cordgrass marsh at slightly higher elevation. The mid-elevation salt marsh was planted by researchers to test experimental assemblages (approximately 5,000 seedlings). These are presented in greater detail under Model Marsh Planting Experiments. The project also salvaged and replanted 0.55 acres of high marsh on a berm that was constructed to protect the site from flooding of Goat Canyon Creek.

The cordgrass zone was planted in three densities to compare rates of vegetative expansion. Cordgrass plugs, salvaged from the northern arm of the estuary, were approximately four to six inches in diameter and were planted on one-meter, two-meter and four-meter centers (Figures 25A and 25B). To assess the effect of soil enrichment on cordgrass survival and growth, six of 12

FIGURE 23. MODEL MARSH

replicated plots of cordgrass were amended with processed kelp mixed with existing soil. Transects were located diagonally across each plot and point intercept and cover class data using 0.25 m² quadrats data were collected to quantitatively assess cordgrass cover. Qualitative monitoring included general assessment of plant health and vigor and photographs of cordgrass through time. Soil salinities at the root zone were recorded during each monitoring period.

Cordgrass monitoring was conducted each spring (March–April) and fall (September–October) representing the beginning and end of the growing season. Monitoring was conducted for three years, 2000–2003. In 2003, light-footed Ridgway's rails were detected in the salt marsh and monitoring was terminated to avoid disturbing this endangered species.

MODEL MARSH PLANTING EXPERIMENTS

The Model Marsh was planted with five native halophytes—salty marsh daisy, California sea lavender (*Limonium californicum*), saltwort, alkali heath and estuary sea-blite. These five species were planted in equal numbers (108 plants per species for a total of 540 plants distributed in 108 experimental plots in dense clusters (10, 30 and 90 centimeters apart). Each cluster had one plant of each of the five species. The vegetation of the Model

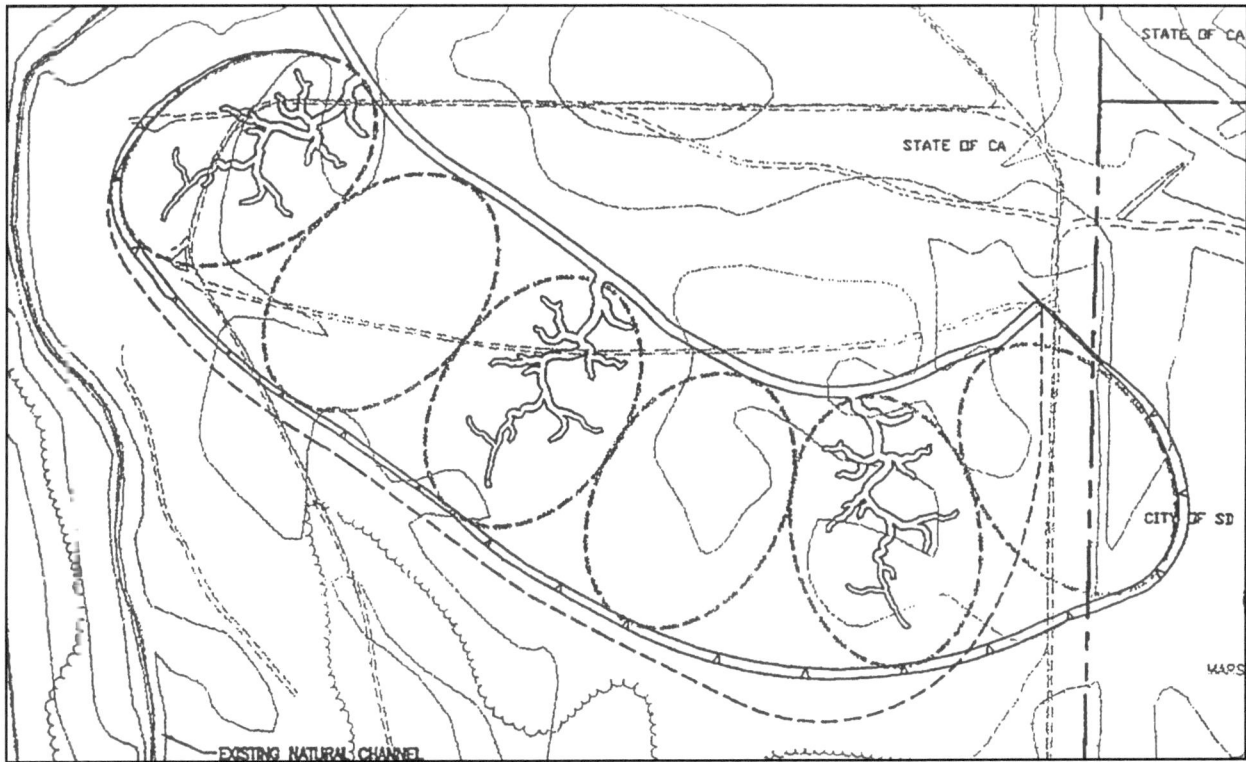

FIGURE 24. LAYOUT OF TIDAL CREEKS

Marsh outside of the seedling plots was assessed using belt transects.

In an attempt to reintroduce Bigelow's pickleweed and estuary sea-blite following restoration of tidal influence in 1984, 2,000 seeds of each species were broadcast in 1998 in an experiment that included plots with and without seeds and with and without removal of the Pacific pickleweed canopy. It was hypothesized that waterlogged depressions in the marsh plain would provide microsites favoring the annual Bigelow's pickleweed over the perennial Pacific pickleweed thereby providing a mechanism where Bigelow's pickleweed could recover after years of decline. Researchers created five- and 10-centimeter deep depressions on

the marsh plain of the Model Marsh using 0.25-m² x 0.15-m cylinders to remove soil and Pacific pickleweed, then removed the desired depth of soil and replanted the Pacific pickleweed and introduced Bigelow's pickleweed seeds to the depressions.

A delay in opening the site to tidal flow occurred when cultural resource monitors encountered evidence of Native American artifacts in the form of a shell midden. Under the law, anthropologists were required to excavate 100 soil pits; however, none showed significant artifacts, only shells, small chards and some charcoal. The 1.5-month delay resulted in the site being opened during mid-February's low-amplitude tides, rather than the spring tides of December and January. As

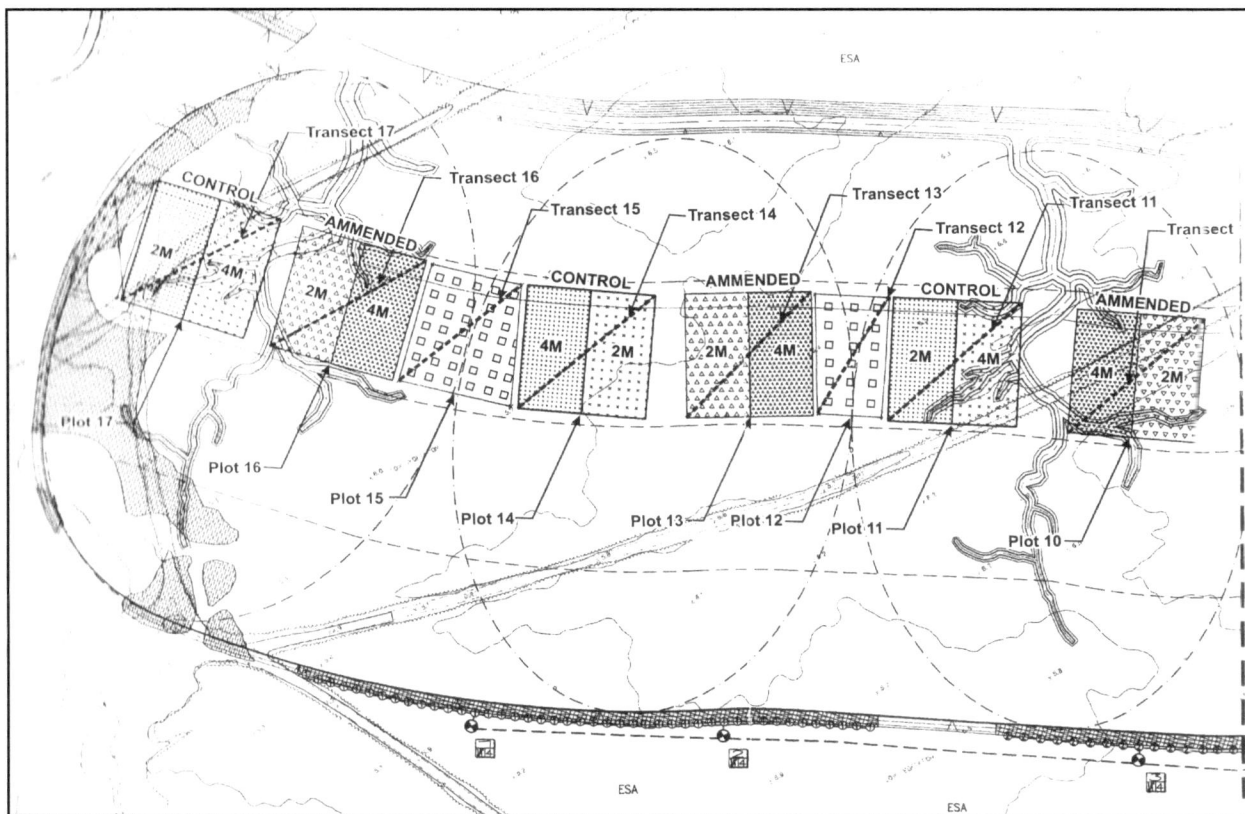

FIGURE 25A. DESIGN OF MODEL MARSH

a result, the excavated marsh plain dried out and became hypersaline. All 5,000+ planted seedlings died. Researchers did not have enough remaining seedlings to reestablish the entire marsh plain, so they designed a new experiment with 180 plants (five-species assemblages introduced with varied cluster density, varied proximity to creeks, and with or without the soil amendment (kelp compost). Monitoring showed that all three treatments increased establishment or growth of at least one species in the assemblage.

A related study assessed seed dispersal and seedling emergence from wrack, rabbit pellets, and soil samples. Tidal dispersal into the Model Marsh occurred mainly in winter, and 90% of emergent seedlings were *S. pacifica* (which was also common in wrack). All seedlings that emerged from rabbit pellets were invasive (*Mesembryanthemum crystallinum* and *Carpobrotus edulis*). Thus, to restore diverse vegetation to bare excavation sites would require seeding or planting.

RESULTS

Amending the soil with processed kelp greatly increased plant growth. Cordgrass canopy cover increased with each survey with a dramatic increase in cover between

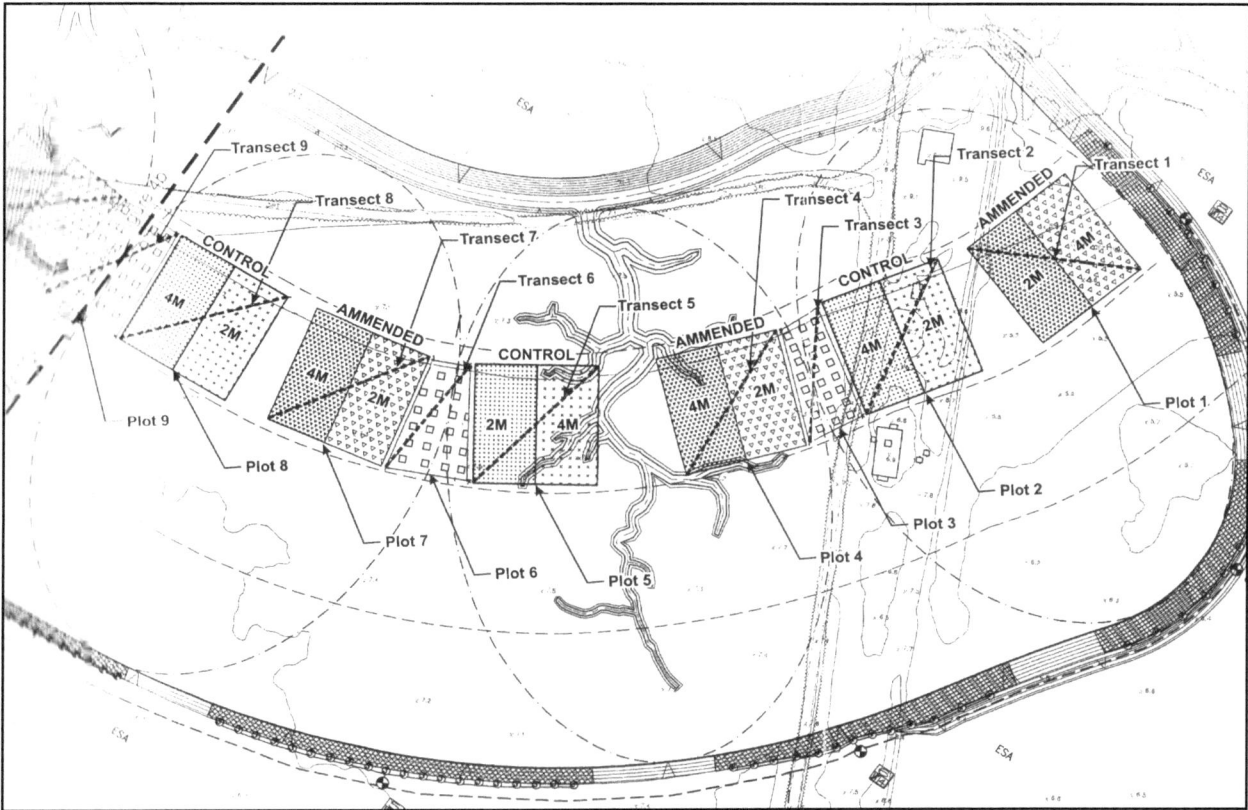

FIGURE 25B. DESIGN OF MODEL MARSH

March 2002 and April 2003 (Figure 26). Cordgrass planted on two-meter centers with soil amendments showed a greater increase in cover over time than unamended plots with cordgrass planted on one-meter centers. Amended plots planted on four-meter centers increased nearly as much as unamended plots planted on one-meter centers (Tierra Environmental Services 2004).

The presence of tidal creeks did not show a corresponding increase in cordgrass expansion. The average point-intercept hits for plots without channels was higher than those with creeks. The monitoring study concluded that soil amendment appeared to have a greater influence on cordgrass growth than did planting densities.

Experimental planting of seedlings of native marsh species in clusters, in proximity to tidal creeks and with kelp amendment increased the growth of at least one species in each assemblage.

Experiments on re-establishing Bigelow's pickleweed showed that the 10-centimeter depressions reduced cover by Pacific pickleweed by approximately 30%; that Bigelow's pickleweed grew taller and produced more flowers in waterlogged

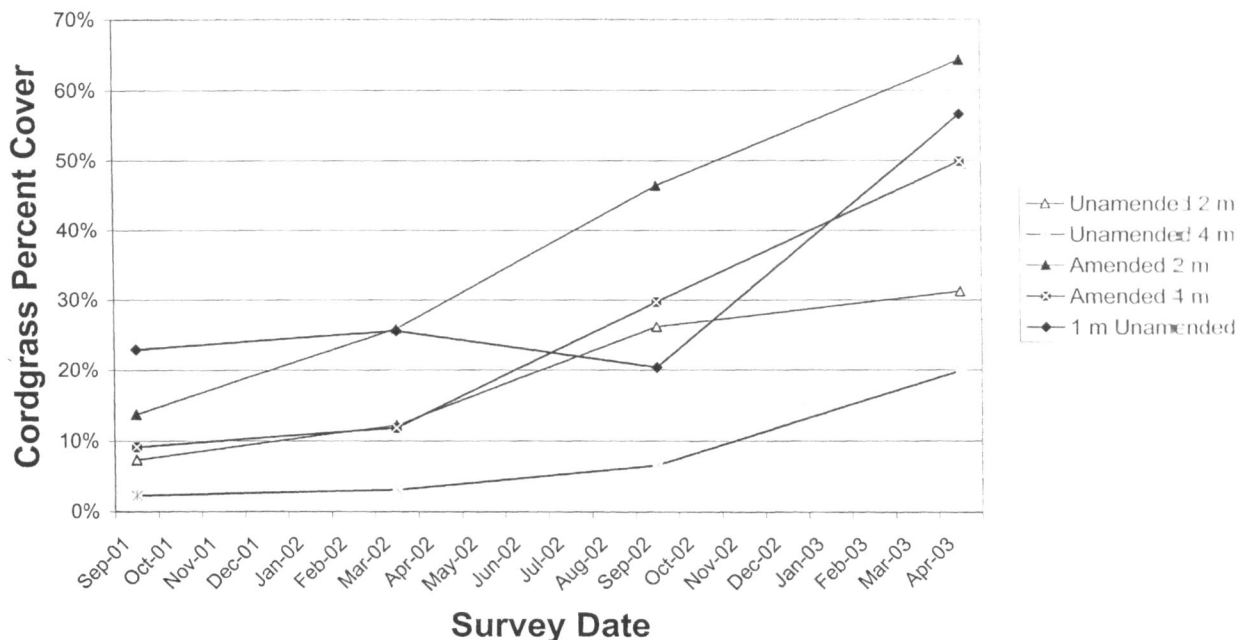

Figure 15. Cordgrass Vegetative Cover for 2 m, 4 m, and 1 m Densities within Unamended and Amended Plots

FIGURE 26. CORDGRASS GROWTH GRAPH

sites with low soil redox potential, and that Bigelow's pickleweed completed its life cycle in the five-centimeter depressions. Experimentally removing Pacific pickleweed in natural shallow depressions also increased survival of Bigelow's pickleweed. Greenhouse experiments determined that rhizosphere oxidation was the mechanism for Bigelow's pickleweed's tolerance of waterlogged conditions as the annual elevated redox potential to a greater degree than did Pacific pickleweed.

It had been hypothesized that tidal creeks would act as "conduits" that would enhance connectivity between subtidal and intertidal habitats and pools formed naturally in the restoration process to serve as microhabitat "oases" for fish. This hypothesis was tested on high tide use of the restored marsh by two regionally dominant fish: California killifish (*Fundulus parvipinnis*) and longjaw mudsucker (*Gillichthys mirabilis*).

Pools provided abundant invertebrate prey and were a preferred microhabitat of California killifish even when the entire marsh was flooded. However, longjaw mudsucker showed no preference for pools: instead, they benefitted from creek banks where they burrowed. Replicate tidal creeks had significantly higher catches of mudsucker but lower catch and feeding rates of killifish.

LESSONS LEARNED

The Model Marsh project imparted several valuable lessons on cordgrass establishment and growth, as well as qualitative observations on the dynamics of a restored salt marsh. The transplanted cordgrass developed an infestation of scale insect (*Haliaspis spartina*) that varied in severity from slight to moderate. Infected areas were carefully monitored over the three-year period to determine if infestation affected survival. Qualitatively, it did not appear to directly affect cordgrass survival or growth. Quantitative analyses supported this observation with cordgrass expanding continually over time and especially in the last year of monitoring.

As predicted, Pacific pickleweed invaded the mid-elevation marsh plain with the first seedlings observed during the second year (May 2001). By April 2003, this species was well established and dominated the mid- marsh plain. Experimental plantings demonstrated that seedlings planted in tight clusters, near creeks and with soil amendments had the greatest rate of survival and growth.

Despite the best of intentions in locating the Model Marsh near the coast (instead of along the Old River Channel), a major flood in November 2004 poured down Goat Canyon and breached the berm, dumping extensive sediment into the restoration site (Wallace et al. 2005). Large areas along the western end (nearest the influx) accumulated up to 30 centimeters (one foot) of sediment; as a result, the intended mudflat became more like the marsh plain, and the cordgrass area lost its direct connection with the tidal channel.

Inflowing sediment from Goat Canyon added to the reworked sediment of the Model Marsh in its initial stage when plants had not established root systems to hold bare sediment in place. The construction of creek networks remained the statistical treatment, but outcomes differed with position in the 20-acre site. Also, new creeks formed between the excavated networks and large pools developed, especially in the eastern cells where the fetch was greatest.

The lesson learned is that excavating third or fourth order tidal creeks might not be necessary. Future research could test the degree to which channels and creeks need to be constructed versus simply started as indentations. Smaller channels would likely form naturally.

The project demonstrated that processed kelp used as a soil amendment had a greater positive effect on cordgrass growth and expansion than did planting density. Processed kelp can increase organic matter, total Kjeldahl nitrogen, inorganic nitrogen, percent soil moisture and decrease bulk density for two or more years after application. It can also increase transplant

survivorship and enhance vegetative growth in the five seedling species and increase stem density and height of California cordgrass. Tight clusters (10-centimeter spacing) can improve seedling transplant survival. This lesson has both ecological and cost implications for restoration projects. From an ecological perspective, the fewer plants that needed to be harvested from an existing marsh for transplant to the restored marsh reduced the damage to the donor site. From a cost perspective, the fewer cordgrass plugs needed to establish the dense cover preferred by species like the light-footed Ridgway's rail, the lower the cost. Unfortunately, processed kelp is no longer available in the San Diego region and an inexpensive replacement has not been identified.

The Model Marsh also demonstrated the nearly exponential growth that cordgrass undergoes, starting with widely-spaced plugs and coalescing to 80% cover and greater in three years. The project was successful in providing habitat for the endangered light-footed clapper rail which prefers dense stands for nesting.

The Oneonta Slough Widening was not implemented due to impacts to sensitive salt marsh from the equipment required to excavate and haul the 26,000 ft^3 of material. The predicted eastward migration of the dune system has not occurred, largely due to active management of the dunes by the USFWS in the north arm.

The 495-acre restoration was not implemented due to impacts associated with the proposed river training berms and cost to dispose of 6,000,000 yd^3 of soil off-site. The project was deemed complete with the attainment of project success criteria following construction of the Connector Channel and Model Marsh.

For more information on restoration experiments conducted at the Model Marsh see:

Morzaria-Luna, H., and J. B. Zedler. 2007. Does seed availability limit plant establishment during salt marsh restoration? Estuaries 30:12–25.

O'Brien, E., and J. B. Zedler. 2006. Accelerating the restoration of vegetation in a southern California salt marsh. Wetlands Ecology and Management 14:269–286.

Wallace, K.J., J.C. Callaway, and J.B. Zedler. 2005. Evolution of tidal creek networks in a high sedimentation environment: A five-year experiment at Tijuana Estuary, California. Estuaries 28:795–811.

Larkin, D.J., S.P. Madon, J.M. West and J.B. Zedler. 2008. Topographic Heterogeneity Influences Fish Use of an Experimentally–Restored Tidal Marsh. Ecological Applications 18(2): 483–496.

Additional Scientific Research Projects Associated with TETRP

The two TETRP projects that were implemented were designed to include restoration research components to further understanding of the physical and biological interactions of the restored sites. These research components were conducted by Dr. Joy Zedler, researchers at PERL, and graduate students at SDSU and the University of Wisconsin and led to a number of publications in scientific journals. In some cases, the results were presented in multiple articles and the projects presented in this book are not an exhaustive list of all publications nor all of the restoration-oriented research conducted at Tijuana Estuary.

Declining Diversity in Natural and Restored Salt Marshes: A 30-Year Study of Tijuana Estuary. Joy B. Zedler and Janelle M. West. 2007. Restoration Ecology 16(2); 249–262.

As indicated in the title, this study integrated historic changes in the natural ecology of Tijuana Estuary with restored sites; specifically, the Oneonta Slough Connector Channel (Tidal Linkage) and the Model Marsh (Friendship Marsh). The study included data on salt marsh vegetation collected along transects in the natural salt marsh from 1974 to 2004 and the responses to the disturbances presented previously in this document, i.e., sedimentation, inlet closure and flooding. These disturbances led to the loss of two short-lived halophytes—Bigelow's pickleweed (*Salicornia bigelovii*) and estuary sea-blite (*Suaeda esteroa*) and dominance by long-lived Pacific pickleweed (*Salicornia pacifica*) and highly productive salt marsh daisy (*Jaumea carnosa*).

Changes in the Natural Marsh

The loss of diversity in the natural salt marsh was attributed to a response to both pulsed and ramp disturbances. Pulse disturbances are defined in this context as discrete events with ephemeral or relatively short-term ecosystem effects and ramp disturbances are defined as cumulative processes leading to long-term changes to ecosystem structure and function. The strongest pulse disturbance was inlet closure, which lasted eight months and coincided with a period of no rainfall. The marsh plain dried, became extremely hypersaline and many plants and animals died. Short-lived annual plant species did not recover.

From 1978 to 2004, flooding was a frequent pulse disturbance. As a result, cumulative sedimentation persisted as a ramp disturbance that elevated the marsh plain and led to increased salinity. Vegetative responses to ramp effects were subtle, in part because salt marsh diversity was already depleted by drought when frequent floods began depositing sediment on the marsh plain. Two productive succulents, Pacific pickleweed and salt marsh daisy, were poised to gain dominance through vegetative expansion and enhanced growth. Through use of the long-term data set and multiple restoration efforts, the authors suggest:

1) how diversity developed in the marsh prior to 1974;
2) how diversity was lost in both the natural marsh and restored marshes;
3) the difficulty in predicting future events; and,
4) how diversity might be restored.

How diversity in the marsh developed prior to 1974. The 1974 data characterized the diversity of the natural salt marsh at Tijuana Estuary, capturing the culmination of a 20-year period preceding 1974 when there were no major river floods or inlet closures. On average, 4.2 species co-occurred at the 0.25 m² scale used to track diversity on the marsh plain. Pacific pickleweed was widespread but not dominant—only the annual Bigelow's pickleweed achieved dominance.

The persistence of Bigelow's pickleweed and estuary sea-blite at Tijuana Estuary suggests that diversity developed in the absence of catastrophic sedimentation. It was hypothesized that these species persist in microhabitats where competition with perennials is suppressed. Microhabitats for estuary sea-blite include channel edges where side lighting could stimulate germination. Microhabitats for Bigelow's pickleweed appear to be shallow depressions in the marsh plain that retain water, as well as sluggishly draining marsh plains.

How diversity was lost in the natural and restored marsh. Five marsh species—Bigelow's pickleweed, estuary sea-blite, alkali heath (*Frankenia salina*), saltwort (*Batis maritima*) and salt marsh daisy—declined rapidly during inlet closure in response to soil desiccation and increased soil salinity. The perennials—saltwort, alkali heath and salt marsh daisy—had recovered by 2004. In contrast, the short-lived Bigelow's pickleweed and estuary sea-blite did not recover. Pacific pickleweed gained dominance during 1984, presumably by growing deep roots that could follow the receding ground water levels. The loss of Bigelow's pickleweed and estuary sea-blite was attributed to shallow growing roots that could track declining water tables.

PREDICTING FUTURE EVENTS

Based on disturbance history, predicting the future of species diversity at Tijuana Estuary was determined to be difficult. Storms and associated sediment deposition occur irregularly in the Mediterranean climate that typifies the southern California coast. Drought and coincidental mouth closure has been proven to be catastrophic to the estuary, yet predictions of either event occurring at a given time or level of intensity are difficult to make with any accuracy. In addition, sea level rise will interact with disturbance sequencing. Current trends in diversity loss might become reversible with sea level rise. Although predicting the future may not be possible, the authors suggest that future restoration efforts include the following:

- Include efforts to keep rare species (Bigelow's pickleweed and estuary sea-blite) from becoming endangered;
- Include natural topographic hetero-geneity (tidal creeks and depressions) to subdue dominant species;
- Over-excavate sites to the elevation of intertidal mudflat to capitalize on sediment deposition;
- Include small islands on the mudflats to be planted with dense, species-rich clusters to trap sediments and expand the marsh plain horizontally.

LESSONS LEARNED

In essence, the entire study is a summary of lessons learned, including lessons learned from long-term monitoring of a natural salt marsh bolstered by lessons learned from long-term experimental manipulation of species diversity in restored salt marshes. It identifies the types of disturbances that have shaped the present salt marsh at Tijuana Estuary and presents recommendations for reversing loss of species diversity, many of which have been incorporated into current restoration planning.

Does Seed Availability Limit Plant Establishment During Salt Marsh Restoration? Hem Nalini Morzaria-Luna and Joy B. Zedler. 2007. Estuaries and Coasts Vol. 30 No. 1, p 12–25.

This study was conducted at the restored Tidal Linkage, the restored Model Marsh and at the natural salt marsh at Tijuana Estuary.

It examined temporal and spatial patterns of seed dispersal and seed bank accumulation on the marsh plain of each site based on seedlings that emerged from floating debris, wrack, rabbit pellets and soil samples in controlled laboratory experiments. Seed dispersal was limited for most marsh plain species. Tidal dispersion occurred mostly in the winter. Seedling density and species richness from germinated tidal debris were highest following high spring tides, with Pacific pickleweed comprising greater than 90% of emergent seedlings. Pacific pickleweed seedlings were also the dominant species in soil samples (up to 63%) and wrack (60%) with other species present at much lower densities. Seed bank accumulation at both restored sites was low with few species abundant. Seedlings that emerged from the Model Marsh site, two years after construction, were dominated by invasive exotics (64%). Seedlings that emerged from the Tidal Linkage, five years after construction, were mostly Pacific pickleweed (63%) and those from the nearby extant marsh were mostly arrow grass (*Triglochin concinna*; 70%), despite a more diverse vegetation assemblage. Seedlings emerging from rabbit pellets were all invasive non-natives, including crystalline ice plant (*Mesembryanthemum crystallinum*) and Hottentot-fig (*Carpobrotus edulis*). Emerging seedlings were much sparser in soil samples from the younger Model Marsh than the older Tidal Linkage and extant marsh. The authors concluded that, because dispersal is limited for most species, restoration of a diverse marsh requires planting or seeding.

This study reinforced earlier observations that Pacific pickleweed is an aggressive colonizer of both natural and restored sites and should not be intentionally planted in order to achieve a more diverse plant assemblage. Planting or seeding of restored sites with species other than pickleweed is necessary to achieve a highly diverse salt marsh.

ADDITIONAL RESTORATION PROJECTS AT TIJUANA ESTUARY

8 TIJUANA ESTUARY—FRIENDSHIP MARSH RESTORATION PROJECT, FEASIBILITY AND DESIGN STUDY

Project Consultants:
Tierra Environmental Services, Rick Engineering Company, Howard Chang Consulting, Jenkins Consulting, AMEC Earth and Environmental, George Mercer, Landscape Architect

Restoration Ecologist:
Chris Nordby, Tierra Environmental Services

Project Administration:
Mayda Winter, SWIA

Date: 2003–2008

Funding Agency:
California State Coastal Conservancy

Reported in:
Tijuana Estuary–Friendship Marsh Restoration Feasibility and Design Study. Tierra Environmental Services. March 2008.

The Tijuana Estuary–Friendship Marsh Restoration Feasibility and Design Study—also known as TETRP II, was a restoration for the south arm of the estuary in the same context as TETRP. A multidisciplinary team composed of restoration ecologists, civil engineers, physical scientists, fluvial and coastal hydrologists, archaeologists and landscape architects developed plans to restore approximately 250 acres of estuarine habitat in the southern estuary, approximately one half the acreage of TETRP.

The project team developed several goals for the restoration project, including:

- Increase tidal prism;
- Restore areas of former salt marsh, tidal channel, and mudflat affected by sedimentation to the maximum extent possible;
- Restore barrier beach and dunes;
- Increase habitat for endangered species;
- Increase area of undisturbed transition zone;
- Build in topographic relief to prevent sudden loss of restored habitat from flood events;
- Incorporate research and adaptive management.

METHODS

Three restoration alternatives were developed and a preferred alternative (Alternative B) that included nearly equal proportions of subtidal, intertidal mudflat and intertidal low and mid-high salt marsh was selected for further analysis (Figure 27; Table 2). A prominent feature of Alternative B was a large subtidal basin, designed to capture mobile sediment already in the system from trans-border canyons and increase the existing tidal prism of the estuary by approximately a factor of two.

Analysis of the fluvial dynamics of the Tijuana River conducted for the study indicated that the restored marsh would be subject to sedimentation and scour from a 10-year flood event and higher events, e.g., the 25-year event. In order to protect the marsh, a berm and weir were proposed extending approximately 8,500 feet from an existing berm located at the eastern end of the estuary to barrier dunes. An approximately seven-foot-high, 700-foot-long weir in the berm would allow floods higher than the 10-year event to enter marsh but at reduced velocities so that no scouring would occur. The weir would exclude all sediment but

TABLE 2. HABITAT TYPES AND ACREAGE CREATED BY EACH RESTORATION ALTERNATIVE

| Alternative | Habitats Created (acres) | | | | | |
	Subtidal	Mudflat	Low Marsh	Mid–High Marsh	Transition Zone	Total
Alternative A	41.06	41.25	61.02	49.69	9.19	202.21
Alternative B	61.15	60.66	60.03	61.32	7.11	250.27
Alternative C	42.94	53.66	60.19	50.83	9.62	217.24

Legend

- Low Marsh (60.03 acres)
- Mid-High Marsh (61.32 acres)
- Mudflat (60.66 acres)
- Subtidal (61.15 acres)
- Transition Zone (7.11 acres)
- Berm

1 inch equals 800 feet
0   200   400   800

RICK
ENGINEERING COMPANY

FIGURE 27.
ALTERNATIVE B (PREFERRED) RESTORATION PLAN AND HABITAT CONFIGURATION

the fine wash load thereby minimizing sediment deposition. Like the river training berm proposed in TETRP, it was anticipated that this berm could be controversial.

Limited analysis of soil grain size and chemistry suggested that some strata in the proposed project area are suitable for beneficial reuse, i.e., beach nourishment, but that those strata are not consistent across the project area. DDT and derivatives were detected in four of six composited surface samples, which would limit placement of these materials in the nearshore environment. Potential disposal options can include a variety of upland reuses, such as industrial fill or berm creation.

In order to assess potential impacts to cultural resources from excavation for restoration, a trenching study was conducted to examine the potential for buried cultural resources. A total of 48 trenches, each 10-meters x one-meter wide and three-meters to four-meters deep were excavated by backhoe. The trenching study revealed two new sites: a Prehistoric shell midden, indicating a Native American food processing camp; and a collection of WWII-era hardware in oil-drenched soil, providing evidence of military use of the area. No other sites, or significant cultural resources, were found in the project area. Therefore, no significant impacts are expected.

RESULTS AND LESSONS LEARNED

Construction would temporarily impact existing biological resources, including nesting habitat of the state-listed endangered Belding's savannah sparrow (*Passerculus sandwichensis beldingi*). The project was considered to be self-mitigating with the acreage of restored habitats more than offsetting impacts to mostly disturbed habitats.

The project would generate approximately 2,500,000 yd^3 of sediment which would need to be placed at an upland site or on the beach. Five phases of construction ranging from 32 to 75 acres were analyzed and cost estimates for each phase were calculated based on five alternative disposal options and combinations thereof, including:

- Nelson Sloan Quarry at Dairy Mart Road;
- Beach and dune replenishment;
- LA 5 offshore dredged material disposal site;
- Hanson El Monte Pit in Lakeside, California;
- Hanson Miramar Landfill.

There was a wide variety of costs depending on disposal site, ranging from approximately $7,800,000 for the 37-acre Phase 4 to $50,000,000 for the 75-acre Phase 3.

In February 2016, SWIA was awarded a grant from the Wildlife Conservation Board to develop the first phase of this project and initial work is underway. Currently, a Science

Advisory Team has been assembled, and this group is helping incorporate lessons learned from past projects. These include careful consideration of the ability of tidal channels to convey enough water for newly-created restoration sites, incorporation of starter channels, and extensive cultural testing.

9 RESTORATION OF THE FORMER MODEL AIRPLANE CLUB FIELD

Lead Agency:
   California State Parks
Contractor:
   Joe Ellis, Marathon Construction
Restoration Ecologist:
   Paul Jorgensen
Date: 1992–2002
Funding Agency:
   California State Parks
Reported in:
   Unpublished Summary by Greg Abbott, CSP Ecologist

During the mid-1980s until about 1990, a radio-controlled model airplane club operated a paved landing field on an approximately five-acre parcel of land immediately west of PERL in the south arm of the estuary (Figure 28). The planes occasionally crashed at PERL, raising safety concerns by the researchers, and a contentious relationship with the club members developed over the years. Requests that the club relocate were denied and a California state assemblyman sided with the club indicating that they could remain as long as they desired. PERL was subsequently moved to the north end of the estuary due to safety concerns from the high volume of illegal immigrants crossing the border at the southern site. Eventually California State Parks brokered a deal with the club, and it was relocated to the floodplain of the Tijuana River near the International Wastewater Treatment Plant.

METHODS AND RESULTS

With the club relocated, the asphalt paving and associated storage structures were removed and Paul Jorgensen, California State Parks biologist, began a phased restoration of the site. Early documents indicate that this restoration was originally planned to be a salt marsh with a hydrological connected to existing channels to the north. The removal of the paving left a shallow that impounded water during rainy years making it ideal for wetland/transition species such as mule-fat.

FIGURE 28. MODEL AIRPLANE FIELD

Winter rains in 1993 deposited sediment in the depression and introduced non-native species, particularly salt cedar (*Tamarix* spp.), to the site. Beginning in 2000, the salt cedar was removed and willow cuttings were planted during the rainy season. In 2001, another flood event deposited more sediment on the site and introduced non-native castor bean (*Ricinus communis*) to the restored area. Removal of castor bean was delayed that year in order to avoid disturbance to nesting American avocets (*Haematopus palliates*) and black-necked stilts (*Himantopus mexicanus*). This delay allowed castor bean to set seed and spread throughout the site. Continued treatment with herbicide has reduced the density of castor bean but it remains along with other non-native species.

LESSONS LEARNED

This project is an example of a small-scale restoration that faced a number of challenges during planning and implementation. Originally planned as a salt marsh restoration, removal of the asphalt landing surface resulted in a depression that impounded rain water. Thus, the restoration was modified to include primarily mule-fat scrub. Continued impoundment of fresh water as well as deposition of sediment introduced non-native plant species the removal of which was hindered by nesting shorebirds. California State Parks biologist Greg Abbott suggested that one person with a backpack sprayer could have treated the invasive plants even during nesting season with minimal disturbance to the birds. He further suggested that restoration projects, regardless of size, include 10-year maintenance and monitoring periods rather than the five-year plans typical of most projects. This protracted monitoring period would allow for flexibility to respond to unexpected events that reshape the landscape and allow time to control invasive species.

FUDS MMRP Project, Border Field State Park, Remedial Investigation/ Feasibility Study

10

Lead Agencies:
> U.S. Army Corps of Engineers

Project Consultants:
> Bristol Environmental Remediation Services

Project Administration:
> U.S. Army Corps of Engineers

Date: 1994 – 2012

Funding Agencies:
> U.S. Department of Defense

Reported in:
> FUDS MMRP Project, JO9CA704401, Border Field State Park Remedial Investigation/Feasibility Study. U.S. Army Corps of Engineers, L.A. District, August 8, 2102.

The FUDS (Formerly Used Defense Sites) MMRP (Military Munitions Response Program) Project, Border Field State Park (BFSP) Remedial Investigation/Feasibility Study (RI/FS) was conducted by the U.S. Department of Defense to determine whether the former use of BFSP as a military training site resulted in remnant exploded or unexploded ordnances or contamination that could be harmful to human and other receptors. The study was conducted by the U.S. Army Corps of Engineers (USACOE) and their contractors, and included stake holder input from CSP and USFWS and review by the California Department of Toxic Substances Control. The project was initiated in 1994 and a Technical Project Planning (TPP) analysis was completed in 2012.

The goals of the TPP were:

- To understand the impact that the presence of MEC (Munitions and Explosives of Concern) has at the site; and,

- To identify appropriate response actions to reduce the risk of MEC that allows for the reasonable public use of the site.

BFSP was used by the military from 1929 to 1961; first as an auxiliary aviation field to the Naval Air Station in Imperial Beach and later in the 1940s as an aircraft gunnery range, machine gun training center, bombing target, air-to-ground gunnery, and emergency landing field. The site was used in the 1950s by the California National Guard on weekends as a launching area for pilotless target drones. In 1961, gunnery training activities were discontinued. In addition, the Imperial Beach Police Department reportedly used the southern portion of the site in the 1960s as a target range. The site was declared surplus by the DoD in 1971 and was transferred to CSP.

There were two primary use areas that were investigated during the study. MRS01 was a 72-acre site used for practice bombing in the 1940s (Figure 29). Three-pound miniature practice bombs were dropped on a target in the middle of this site. MRS02 consisted of 249 land acres and 2,517 off-shore acres (Figure 29). Five or six Jeep moving targets were strafed by plane and land-based machine guns. The predominant ammunitions used by the Navy for this training were 0.5 caliber and 0.3 caliber machine gun bullets. An Inventory Project Report completed in 1994 detected no munitions and MEC or Munitions Debris (MD). The CSP ranger at the time reported that during scrap metal surveys conducted in the 1970s, only 0.5 caliber munitions were recovered. An Archive Search Report conducted in 2003 confirmed the absence of MEC and noted that three to four feet of sediment had been deposited over much of the site.

A Site Inspection of MRS01 and MRS02 conducted in 2007 included evaluation of 7.65 linear miles of the site, collection and inspection of 11 soil samples for munitions, explosives and metals, and collection of two surface water samples at MRS02 for analysis of munitions, explosives, metals and perchlorate. During this 2007 inspection, there was no evidence of MEC or MD at MRS01 nor any visible evidence of practice bomb usage. Small arms MD (0.3 and 0.5 caliber) were observed at MRS02. No other MD or MEC were observed at MRS02. No explosives were detected in the soil and no munitions-associated metals were detected above background levels were detected in soil at MRS01. No explosives were detected in the soil at MRS02. Seven munitions-associated metals were detected above background levels in the surface soils at MRS02—aluminum, barium, copper, lead, molybdenum, strontium and zinc. Based on the results of a screening level risk assessment, it was determined that there was the potential for unacceptable risk to human health from molybdenum levels in surface water at MRS02 and the potential for ecological risks due to exposure to molybdenum, strontium and copper. Based on these results, it was recommended that the analysis of MRS01 and MRS02 proceed to the RI/FS. The goals of the RI/FS were:

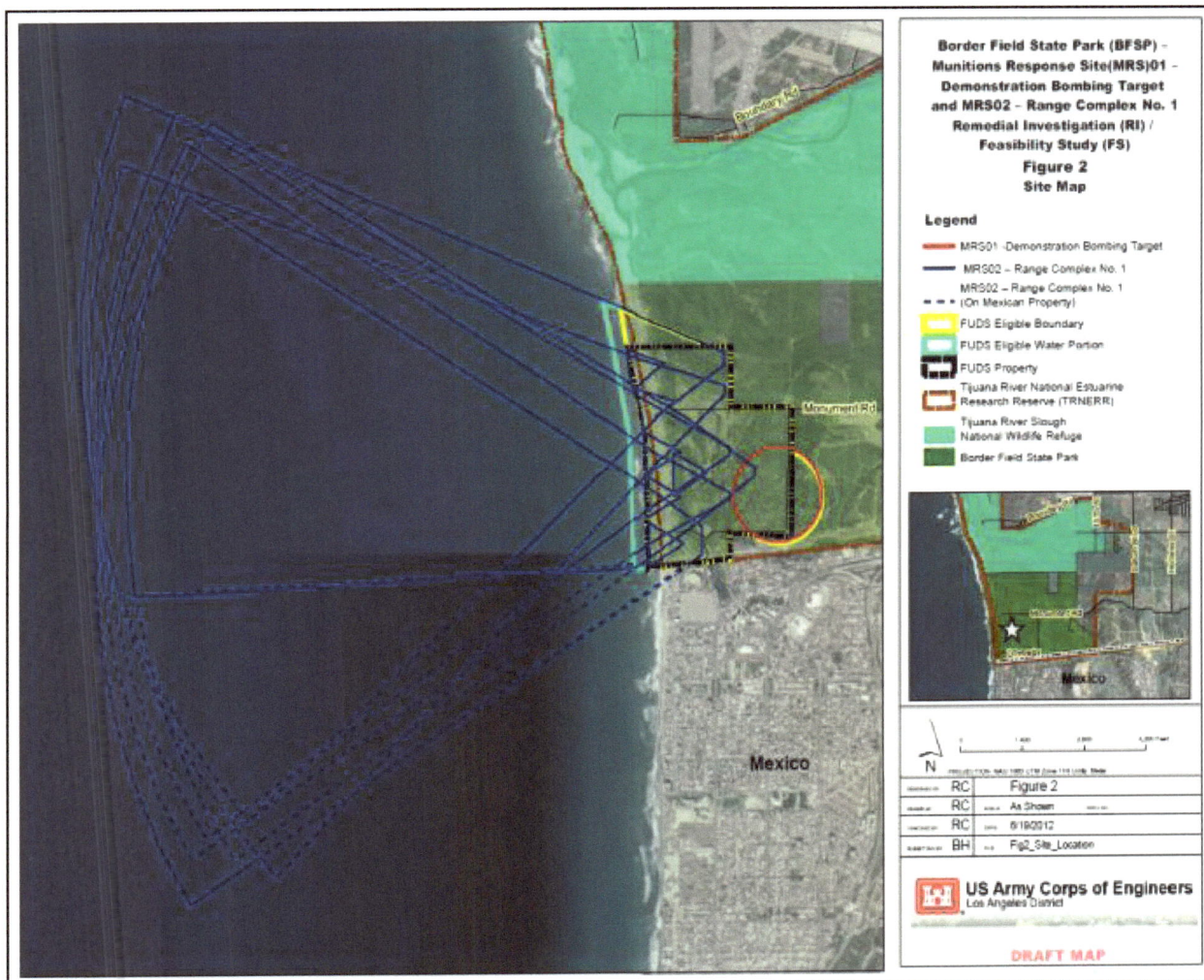

FIGURE 29. MRS01 AND MRS02

- Identify and characterize potential MEC and MC (Munitions Constituents);
- Identify potential hazards and risks associated with remnant MEC and MC;
- Identify, evaluate and present response alternatives for the remnant MEC and MC;
- Develop and screen alternative remedial alternatives;
- Provide detailed cost estimates;
- Provide a comparison of the advantages and disadvantages of each alternative.

Accordingly, a RI/FS strategy, including data collection and analysis, was developed for the project. Remedial alternatives identified in the plan included the No Action Alternative; an Institutional Controls Alternative (signs, training, education, deed restrictions); UXO (Unexploded Ordnance) Construction Support Alternative; Surface Removal Alternative; Subsurface Removal

Alternative; and any combination of these. The preliminary schedule developed for the RI/FS identified February 2013 as the start of field data collection.

The results of the RI/FS were not available for this document. Depending on the results of soil and water sampling and analysis any potential remediation alternative could both inform and affect the proposed 250-acre restoration envisioned by the Tijuana Estuary–Friendship Marsh Tidal Restoration Project. Information on munitions and metals in soil and water could inform the project while alternatives that include surface and subsurface removal could aid the implementation of the restoration project. For these reasons, the FUDS MMRP Project is considered important to the overall restoration effort at Tijuana Estuary and will likely provide lessons learned that are directly applicable to those efforts.

11 SOUTH BAY WATER RECLAMATION PLANT AND THE DAIRY MART ROAD IMPROVEMENTS PROJECT

Project Engineers:
City of San Diego, Metropolitan Wastewater Department (MWWD)
Restoration Ecologist:
Chris Nordby, Tierra Environmental Services,
Mooney & Associates, Merkel & Associates
Date: 1996 – 2006
Reported in:
Environmental Impact Report/Environmental Assessment for the South Bay Water Reclamation Pland and Dairy Mart Bridge and Road Improvements. Appendix B – Biological Resources Technical Report. Tierra Environmental Services. March 13, 1996.

Second Annual Revegetation Monitoring Report for the Dairy Mart Road Revegetation Area. Mooney & Associates. 2001.

60-Month Monitoring Report – South Bay Reclamation Sewer and Pump Station. Merkel & Associates. October 26, 2006.

The South Bay Water Reclamation Plant (SBWRP) and Dairy Mart Road Improvement Project was a City of San Diego, MWWD project consisting of two separate components: the SBWRP and the improved Dairy Mart Road and Bridge, designed to provide access to the SBWRP during the modeled 100-year flood of the Tijuana River.

METHODS

The SBWRP site consisted of a 22.3-acre parcel, located southeast of the intersection of Dairy Mart Road and Monument Road. A Biological Resources Technical Report and project EIR/EA were prepared for the project that examined the biological resource of the SBWRP and four alternative alignments of Dairy Mart Road and Bridge. The Dairy Mart Road and Bridge alignment also consisted of several components:

1) the realignment and improvement of Dairy Mart Road;
2) the replacement of the existing Dairy Mart Road Bridge;
3) the realignment and improvement of the south levee of the Tijuana River channel located between the International Wastewater Treatment Plant (IWTP) access road and the southeast base of the replacement bridge; and
4) removal of the existing Dairy Mart Road and Bridge.

The Biological Resources Technical Report was completed in 1996 and the SBWRP opened in 2002.

The SBWRP site was composed of primarily ruderal fields (21.8 acres) with some disturbed areas (0.5 acre). The road alignment alternatives included several habitat types of different acreages and assemblages, including open water, coastal freshwater marsh, southern cottonwood-riparian forest, southern willow scrub, disturbed floodplain, coastal sage scrub, ruderal habitat, former agricultural land and developed areas. All road alignments impacted USACE, RWQCB and CDFW jurisdictional habitats including both wetlands and non-wetland waters. In addition, riparian habitats, including southern willow scrub and southern cottonwood-riparian forest were occupied by five to six pairs of endangered least Bell's vireo and one pair of nesting light-footed Ridgway's rail utilizing freshwater marsh habitat.

A conceptual restoration plan as mitigation for impacts to sensitive resources, including riparian habitats, jurisdictional habitats and coastal sage scrub was developed. Mitigation ratios ranged from 1:1 for impacts to open water to 3:1 for impacts to high quality riparian and freshwater marsh habitats.

Mitigation for the impacts to sensitive habitats associated with Dairy Mart Road and Bridge realignment/replacement was implemented on an approximately 10-acre site adjacent to the new bridge and consisted of 6.73 acres of southern willow scrub and 3.18 acres of waters of the U.S. The site was planted in 1999 by Merkel & Associates (Merkel) and was monitored for three years by Mooney & Associates (Mooney). An additional off-site mitigation area consisting of 3.6 acres of wetland habitats was required for impacts associated with the treatment plant. This site was planted in 2001 by Merkel and was monitored for five years by Merkel and Mooney.

Results

6.73-acre Mitigation Site

The 6.73-acre site was graded to an elevation that was to within five to seven feet of the summer ground water level and two to three feet above low flow water surface elevation in the Tijuana River. An irrigation system was installed after the site was graded and prior to planting. The southern willow scrub species were planted in June 1999 with 2,661 primarily one-gallon container stock tree species, dominated by arroyo willow (*Salix lasiolepis;* 1,406) and narrow-leaved willow (*Salix exigua;* 903). These were augmented with 1,000 willow cuttings of mule-fat. In addition, the site was seeded with selected riparian herbaceous and shrub species.

Monitoring of this mitigation site was initiated in August 1999 after a 60-day plant establishment period. Twelve 30-meter long belt transects were established; percent cover determined using the point intercept method; and species composition and survival of planted species were determined within a three-meter wide belt.

The results of the Year 3 monitoring event established that the site had achieved its goal of a multi-layered willow riparian community and mule-fat scrubland. The irrigation had been discontinued in fall of 1999 and percent cover was greater than 90%. Natural recruitment of both willows and mule-fat could not be differentiated from planted cuttings or container stock. The endangered least Bell's vireo (*Vireo bellii pusillus*) recolonized the site and the permitting agencies accepted an early release of the project from its proposed five-year monitoring program.

3.6-acre Mitigation Site

The 3.6-acre site was graded to an unspecified elevation and an irrigation system was installed prior to planting, which was completed November 5, 2001. Habitats planted included southern willow scrub, mule-fat scrub and *Baccharis* scrub. The final monitoring plan did not specify the quantities of individual species and the conceptual restoration plan was not available for this summary.

Monitoring of the 3.6-acre site was initiated in November 2001 after a 120-day plant establishment period. Thirteen 20-m long transects were established in the three habitat zones and percent cover was determined using the point intercept method. In addition, average plant heights were measured for selected species.

The monitoring program was conducted for the full five-year period. The final monitoring event demonstrated that mean percent cover was 80%, 10% short of the project goal of 90%; however, the site exceeded the average height requirement of 10 feet and demonstrated natural recruitment of target species. Additionally, two vocalizing male least Bell vireos indicated that the 3.6-acre site was attractive to this endangered species. The report recommended additional plantings in relatively bare areas and postponing final sign-off until spring of 2007. The site received final sign-off in 2008.

LESSONS LEARNED

The principal lesson learned from the 6.73-acre site was that achieving the proper elevation relative to ground water generally ensures a high survival rate of planted species and a high degree of natural recruitment of desired species. An interview with the project manager at the City of San Diego suggested that the site was overplanted and that a reduced planting density would have sufficed. Furthermore, the irrigation system was only employed from June 1999 to October 1999. While supplemental irrigation may have been critical for the establishment of some species, natural recruitment of willows and mule-fat suggests that this system could have been reduced or eliminated. The success of both sites in attracting least Bell's vireo, as well as other sensitive bird species, suggests that mitigation sites can mimic natural riparian wetland habitats, at least in the short term. Least Bell's vireo prefer dense, relatively short willow and mule-fat thickets adjacent to open areas as breeding habitat. As much of the riparian habitat in the Tijuana River is mature willow gallery forest, less mature sites offer additional areas for recovery of this species. Unfortunately, much of the willow habitat in the Tijuana River Valley has been killed recently by the *Fusarium euwallacea* die-back associated with the shot hole borer beetles.

12 HOLLISTER STREET BAILEY BRIDGE REPLACEMENT PROJECT

Design Engineers:
Simon Wong Engineers
Restoration Ecologist:
Chris Nordby, Tierra Environmental Services
Landscaping Contractor:
Habitat West
Date: 1997–2002
Funding Agency:
City of San Diego, Engineering and Capital Projects
Project Administration:
Brad Johnson, City of San Diego
Reported in:
Third Annual Monitoring Report for the Hollister Street Bridge Replacement Project Mitigation Program. Tierra Environmental Services. 2002.

In 1993, winter storms resulted in flooding that diverted the Tijuana River at Hollister Street resulting in the creation of a new channel to the north of the existing channel (Figure 30). The diversion was the result of the unpermitted fill on the west side of the bridge known as the "Brown Fill" named after the former landowner. Despite repeated warnings from regulatory agencies, including the USACE, Brown continued to place fill into the Tijuana River creating a wedge of land that restricted flows (Figure 30). During the 1993 rainy season the flooding Tijuana River was diverted to the north by this fill and created a new channel to the estuary. The existing bridge over the river was damaged and subsequently demolished. A temporary "Bailey" bridge was erected to allow vehicular access. The City of San Diego developed and implemented plans for replacing that temporary structure with a permanent bridge. The bridge was constructed in winter of 1997.

Construction resulted in impacts to 0.6 acre of southern willow scrub wetland, 0.1 acre of freshwater marsh; and 0.17 acre of non-wetland waters. Required mitigation included using 0.6 acre of excess willow and mule-fat habitat that was previously planted in the Tijuana River Pilot Channel Mitigation Site; removing 1.65 acres of exotic vegetation including salt cedar (*Tamarix* spp.), castor bean (*Ricinus communis*), giant reed (*Arundo donax*), and tree tobacco (*Nicotiana glauca*) within the floodplain and along the north bank of the Tijuana River northern flow channel, west of Hollister Street Bridge; grading down 0.6 acre of disturbed upland

habitat immediately adjacent to the bridge construction site in the Tijuana River to the elevation of the existing river bed; and planting willow cuttings under the rip-rap that was placed for slope protection along the sides of the river channel banks.

METHODS

Initial exotic eradication was conducted in August 1998 and consisted of cutting each stalk or stem and immediately applying water-safe herbicide (Garlon 4). Above-ground biomass was removed from the project site after cutting. Follow-up eradication efforts were conducted in October 1998 with minor herbicide application the first week of March 1999. Inspection of the site in April 1999 revealed that all target species had been successfully treated.

The willow wattle bundles and cuttings were constructed and installed in March 1999. Wattle bundles were constructed by selectively pruning willow species in the project area taking care not to harvest more than 10%–15% of the suitable branches from any given tree. Cuttings were roughly six feet in length and no less than 0.25 inch in diameter. Cuttings were collected from three willow species: black willow (*Salix gooddingii*), arroyo willow (*Salix lasiolepis*) and narrow-leaved willow (*Salix exigua*). Wattles consisted of groups of willow cuttings approximately eight inches in diameter bound by biodegradable twine. A series of these eight-inch-diameter bundles were laid end-to-end in two trenches excavated at the toe of the rip-rap slope and approximately

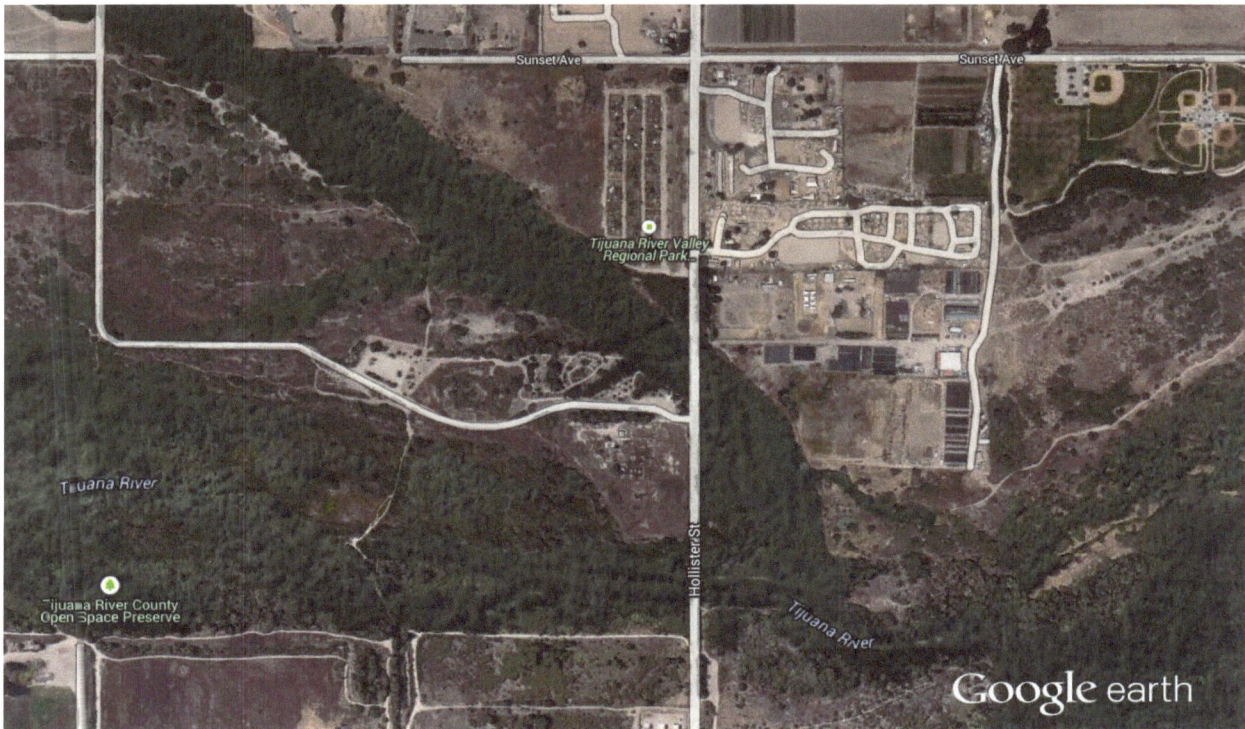

FIGURE 30.
AERIAL PHOTOGRAPH SHOWING BROWN FILL AND NORTHERN CHANNEL

three feet upslope. The wattle bundles were then covered with a shallow (one to two inches) layer of dirt to minimize drying. The buried wattle bundles were irrigated for approximately 20 minutes three times per week.

RESULTS

This method of propagating willows was very successful. All three species of willow developed aerial growth and roots. With the exception of a small area located on the southwest bank where irrigation water apparently failed to reach, the two rows of willows grew rapidly. A small-scale experiment was undertaken by the City's Project Engineer, John Revels, to determine if fewer willow cuttings could produce a similar result. Mr. Revels buried individual cuttings versus eight-inch diameter bundles of cuttings, thereby reducing labor and impact on the source of cuttings. Similar results were obtained using this method. It was recommended that future restoration efforts utilize this modified method.

Cuttings installed in the floodplain of the river consisted of willow branches approximately four to five feet in length and no less than 0.25 inch in diameter pruned from mature trees in the project area. Each cutting had at least two side branches, or buds, and was buried vertically in the sandy

substrate to a depth of about two feet leaving approximately two to three feet extending above the floodplain. Cuttings were taken from each of the three species collected for wattle bundles; black, arroyo and sandbar willow. In all, 580 willow cuttings were installed on approximately two-foot centers using this method.

Survival for willow cuttings on the floodplain was moderate, despite the use of supplemental irrigation. A total of 249 cuttings, or 43% of the total planted, survived.

A unique aspect of this project involved the planting of non-native salt cedar (*Tamarix aphylla*) as mitigation for removal of the same species during construction. These trees were identified by a local ornithologist as an important source of food (insects) for migrating song birds as they arrived in the river valley in early spring. Despite City and County of San Diego policies restricting the intentional planting of invasive, non-native species and the project goal of eliminating salt cedar species in the project area, 10 individual salt cedar were planted along with eight Fremont's cottonwood trees (*Populus fremontii*). All survived through the postconstruction monitoring period.

Project permit conditions stipulated that postconstruction monitoring be conducted for five years. The project, however, met the success criteria in three years and monitoring was terminated at that time.

LESSONS LEARNED

The Hollister Street Bailey Bridge Replacement Project imparted several lessons that can inform future restoration projects, as summarized here.

- Burial of willow wattles in the dirt cap of rip-rap slopes is an effective method for revegetating armored slopes;
- Burial of individual cuttings can accomplish the same results with reduced labor and cost;
- Treatment of non-natives with Garlon 4 (Triclopyr) is an effective method for their control;
- Revegetation using non-native species should be discouraged. Despite providing a food source for early migratory bird species, salt cedar should be replaced with native species, such as cottonwoods, that can provide the same or a similar food source. This is now the prevailing management practice in the River Valley.

13 NAPOLITANO RESTORATION PROJECT

Prime Contractor:
> Brown and Caldwell Engineers

Restoration Ecologist:
> Chris Nordby, Tierra Environmental Services

Date: 1998–2001

Funding Agency:
> Caltrans District 11

Reported in:
> Tijuana Estuary Wetlands Restoration Project Draft Removal and Restoration Plan. Prepared for California Department of Transportation District 11. Brown and Caldwell and Tierra Environmental Services. 1998.
>
> Tijuana Estuary Wetlands Restoration Project, Monitoring and Maintenance Report Year 3. Prepared for California Department of Transportation District 11. Brown and Caldwell and Tierra Environmental Services. 2001.

The Napolitano Restoration Project entailed the removal of approximately 9,000 cubic yards of fill deposited in the intertidal salt marsh to create a pad for a residential development in the northern arm of Tijuana Estuary adjacent to Seacoast Drive (Figure 31). The 1.25-acre fill was created in the 1970s under a grading permit issued by the City of Imperial Beach and had been used as an unofficial parking lot by residents of Imperial Beach. It was restored by Caltrans as a Supplemental Environmental Project as part of a settlement agreement with the U.S. Environmental Protection Agency, the Natural Resources Defense Council, and the San Diego Baykeepers regarding the discharge of Caltrans storm drains into San Diego Bay. Caltrans acquired the property from the Napolitano Trust and developed and implemented a wetland restoration plan.

METHODS

The restoration plan included removal of fill to match the elevation of the adjacent salt marsh and excavation of a tidal channel linking to existing tidal channels to the south and east. The plan was designed to mimic the plant species composition of a reference marsh located immediately south of the fill. The reference marsh was a mixture of low- and mid-marsh species dominated by California cordgrass and Pacific pickleweed with subdominants including saltwort, alkali heath, sea lavender and salt marsh daisy. Accordingly, the site was planted with 2,700 plugs of California cordgrass harvested from the northeastern edge of the former sewage lagoon at the estuary and an approximately equal number of Pacific pickleweed grown from seed and cuttings. Both were planted on one-meter centers. The four subdominants were planted on 2.5-meter centers at quantities of 205 each.

A narrow transition zone was planted with 201 individuals of saltgrass (*Disitchlis spicata*) grown from cuttings in one-gallon containers as well as 18 individual Parish's pickleweed (*Arthrocnemum subterminale*) salvaged from the site and eight boxthorn (*Lycium brevipes*) also salvaged.

Project specifications dictated that 80% of the planted species survive the first year with 100% surviving thereafter. To insure survival criteria were met, the site was substantially overplanted. The original planting plan specified a total of 5,385 plants. To allow for some mortality during plant establishment, an additional 9% was added to the nursery contract specifications for a total of 5,864 plants to be delivered. However, in an attempt to compensate for the low quality of plant material (plants were undersized and immature), the nursery delivered an additional 865 plants. An additional 794 cordgrass plugs were also planted to maximize this species habitat type. In total, an additional 1,659 plants were planted relative to the approved planting plan. The final planting plan of 7,523 individuals represented an overplanting of approximately 37%. Thus, the site could

endure substantial mortality and still meet survival criteria.

During the first day of excavation, an approximately one-foot thick layer of cobbles and asphalt waste was discovered at the elevation of the proposed marsh plain. The asphalt was excavated and disposed of at an appropriate landfill and the void was filled with a 2:1 mixture of topsoil and processed kelp. This nutrient rich mixture accelerated plant growth following initial establishment.

Monitoring of the project was conducted quarterly during Year 1 and semi-annually in Years 2 and 3 and included qualitative and quantitative methodology. Qualitative monitoring consisted of documentation of plant health and vigor; overall site conditions; and photographic documentation of marsh development from permanently established stations. Quantitative monitoring included assessment of plant cover along permanently established transects using the point intercept method at 0.5 meter intervals. To assess plant survivorship, live individuals were counted within a five-meter wide belt transect centered on the point intercept transect. Cover data were also collected using 0.25-m² quadrats along each transect at five-meter intervals. Six 50-meter-long transects were monitored within the restored marsh and the results compared to three reference transects immediately south of the project in the natural salt marsh. Although monitoring was originally scheduled for three years, the presence of the endangered light-footed Ridgway's rail in Year 3 resulted in the

FIGURE 31. NAPOLITANO SITE

termination of monitoring at the request of the USFWS Refuge.

RESULTS

The restored marsh developed rapidly, with areas dominated by Pacific pickleweed demonstrating a mean of 87% cover by the end of Year 2 (n = 3 transects). Areas dominated by cordgrass had a mean cover of 69% (n = 2 transects) by the end of Year 2 and the single transect located in high marsh dominated by saltgrass demonstrated 71% cover. These coverage data exceeded the coverage data at the reference transects by the end of Year 2. The topsoil/processed

kelp amendment and the intentional overplanting of salt species was attributed to this early achievement of success criteria.

LESSONS LEARNED

Restoration of the 1.25-acre Napolitano site to intertidal wetland habitat imparted a number of important lessons. Perhaps the most important lesson is that subsurface investigation of soil characteristics should extend below the proposed surface elevation of the restored site. At the Napolitano site, 20 subsurface soil cores were collected at a depth of six feet below existing grade. The cores were subsequently analyzed for

potential contaminants and grain size. No contaminants were detected, yet a one-foot deep layer of contaminated material was discovered at the level of the marsh plain. Extending the soil cores to eight feet below surface would likely have detected the asphalt layer prior to construction, allowing for the development of a disposal and fill plan during the project permitting phase.

The site was substantially overplanted due to the poor condition of nursery-grown plants. Although Caltrans would have been within the contract rights to reject the plants, rejection would obviously not accomplish the goal of planting the marsh within the time frame of the project. Therefore, the decision was made to accept the plants, including the additional quantities offered by the nursery, as partial compensation, and to plant all plants.

It is unlikely that overplanting at the densities that were done for this project are necessary for meeting either initial survival criteria or canopy development criteria. However, should a mitigating entity choose to reduce post-project monitoring costs by increasing installation costs, this project provides evidence that that strategy may be achieved.

Soil augmentation with processed kelp is expensive and is not practical on a large scale. In 1998, processed kelp from the American Kelp Company was $22/yd^3 and topsoil was around $5/yd^3. Furthermore, The American Kelp Company closed its San Diego operations in 2008 and the processed kelp is no longer available locally.

The success of the topsoil augmented with processed kelp presaged an experiment included in the Model Marsh project where replicate 100-foot x 100-foot plots of native soil and kelp mixed in a similar manner were planted with various densities of cordgrass. The results indicated that such a mixture increased development of the cordgrass canopy (see Lessons Learned Model Marsh).

Another lesson learned by the Napolitano project had to do with the proposed plant palette. The restoration ecologist proposed a mix of species present in the adjacent salt marsh, including cordgrass, Pacific pickleweed and a mixture of mid- and high salt marsh succulents and grasses. Some reviewers expressed doubt that this mix of species would be resilient in the long term, predicting that the site would become dominated by pickleweed. That did not occur and the site remains a mixture of cordgrass, pickleweed and other species. While planting with pickleweed is not encouraged at most sites within Tijuana Estuary and other regional wetlands, it was felt that the unique hydrological conditions of this site, i.e., waterlogged, would support such a species mix. When planning a restoration project, the existing or adjacent site conditions should be considered carefully.

14 GOAT CANYON ENHANCEMENT PROJECT

Lead Agency:
> California State Parks

Project Consultants:
> Rick Engineering Company/Burkhart Environmental Consulting/EDAW, Inc.

Restoration Ecologist:
> Chris Nordby, Tierra Environmental Services

Project Administration:
> Mayda Winter, SWIA

Date: 1998–2009

Funding Agency:
> California State Coastal Conservancy

Reported in:
> Biological Assessment for the Goat Canyon Enhancement Project. Tierra Environmental Services. August 2002.

> Third Annual Monitoring Report for the Border Field State Parks Sediment Basins and Road Realignment Mitigation, Monitoring and Maintenance. Prepared for California State Parks. EDAW, Inc. April 2009.

The Goat Canyon Enhancement Project was undertaken to address sedimentation impacts in the southern arm of Tijuana Estuary. The project was a joint effort by California State Parks, the State Coastal Conservancy and SWIA. The project is located in Border Field State Park. The project constructed a series of sedimentation basins located adjacent to Goat Canyon Creek to capture sediment and debris thereby minimizing the amount of sediment that reaches the wetland habitats of the TENERR (Figure 32). In addition, the project proposed to raise the elevation of Monument Road to provide year-round access to Monument Mesa and other parts of Border Field State Park. Prior to construction of the sedimentation basins, water and sediment flowed along Monument Road resulting in frequent closure.

METHODS

The project originally proposed a series of in-line sedimentation basins within Goat Canyon Creek. However, this design resulted in impacts to sensitive vegetation communities, including nesting habitat of the federal-listed and state-listed endangered least Bell's vireo (*Vireo bellii pusillus*). In order to minimize impacts to these habitats, the sedimentation basins were moved to a

disturbed site at the mouth of Goat Canyon that had been used as the stockpile site for sediment excavated from the Model Marsh project prior to it being used to restore Fenton Quarry. The basins were designed to capture approximately 60,000 cubic yards of sediment annually. Despite moving the basins to the disturbed site, there were impacts to sensitive habitats and a substantial mitigation program, including wetland and upland habitats, was required. Impacts to vegetation and federally-listed species, along with proposed mitigation measures, was documented in a Biological Assessment prepared for the project pursuant to Section 7 of the Federal Endangered Species Act (Tierra Environmental Services 2002).

The sedimentation basins were constructed in 2004–2005 after delays in processing the Section 7 consultation. A storm occurred in 2005 prior to completion of the basins and sediment passed through the partially completed basins and was conveyed to the salt marsh downstream resulting in the burial of approximately 20 acres of marsh and salt pan.

The proposed improvements to Monument Road were not implemented based on the determination by the Coastal Commission that impacts to wetlands associated with transportation projects were not allowable under the Coastal Act. Monument Road continues to be susceptible to flooding and closure.

The mitigation site for the Goat Canyon Enhancement Project totaled 25.76 acres, including mule-fat scrub (20.71 acres), southern willow scrub (2.59 acres), maritime succulent scrub (2.16 acres) and southern mixed chaparral (0.3 acre; Figure 33). Supplemental irrigation was provided using a temporary PVC distribution system with long-range overhead sprinklers. The mitigation areas included approximately 27,000 container plants for the wetland areas and 1,200 plants for the maritime succulent scrub visual screening berm. After planting habitat-specific seed mixes were applied to wetland and upland site. California State Parks accepted the 120-day plant establishment period in September 2005 and a five-year monitoring period was initiated.

RESULTS

Monitoring of the mitigation site was conducted by EDAW (now AECOM) and consisted of quarterly monitoring of restored habitats and annual monitoring of the state-listed and federally-listed endangered least Bell's vireo. In addition, groundwater levels were monitored. During year three of monitoring, EDAW determined that the mitigation site had met its five-year success criteria and requested early sign-off by the permitting agencies. The primary year five success criterion was to achieve at least 85% native cover of herbaceous and shrub species. After three years, the southern willow scrub and mule-fat scrub areas had attained 124% and 151.7% native cover, respectively. In addition, in wetland areas, seed mixes germinated well and became established and container plant survival was 95%.

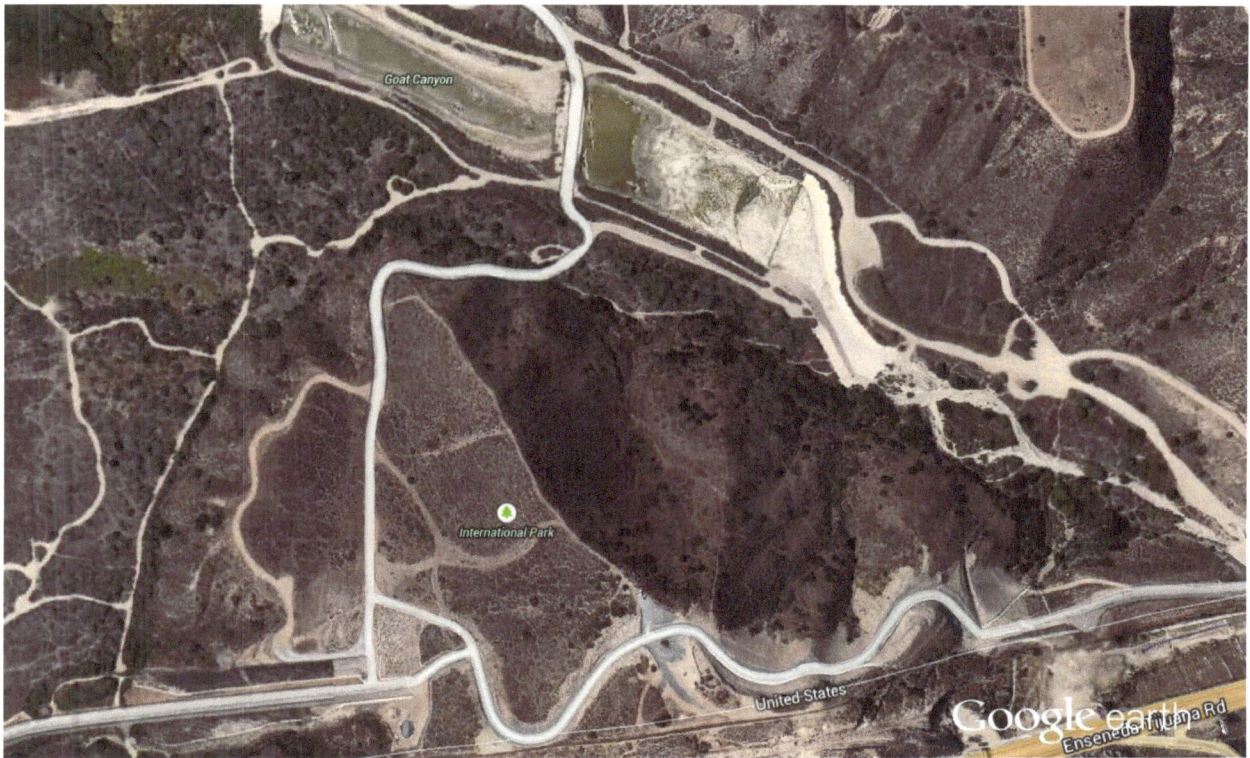

FIGURE 32. GOAT CANYON SEDIMENT BASINS

Due to the positive establishment of these species, the temporary irrigation system was discontinued for over 80% of the site in May 2007. Quarterly monitoring of ground water levels indicated that ground water occurred from 1.2 feet to 11.65 feet below ground surface suggesting that ground water was important to wetland habitats along with periodic surface flows.

In year three, vireos began using the mitigation area for nesting. Prior to that year, all nesting activity occurred in areas adjacent to the mitigation area.

The 2005 winter was one of the wettest winters in history for the San Diego region.

Winter storms recharged ground water aquifers and undoubtedly contributed to the early success of the restoration project. Since that time, San Diego, like much of the arid west, has experienced moderate to severe drought conditions. In some areas of the mitigation site, particularly the southern portion of the mule-fat scrub area, these conditions have led to a gradual decline in the density and vigor of the target species (C. Peregrin, CSP, pers. comm.). While such declines are common with similar naturally established vegetation communities, they can be more evident in mitigation sites that receive additional scrutiny. This site also has been attacked by the Kuroshio shot hole borer, contributing to willow mortality.

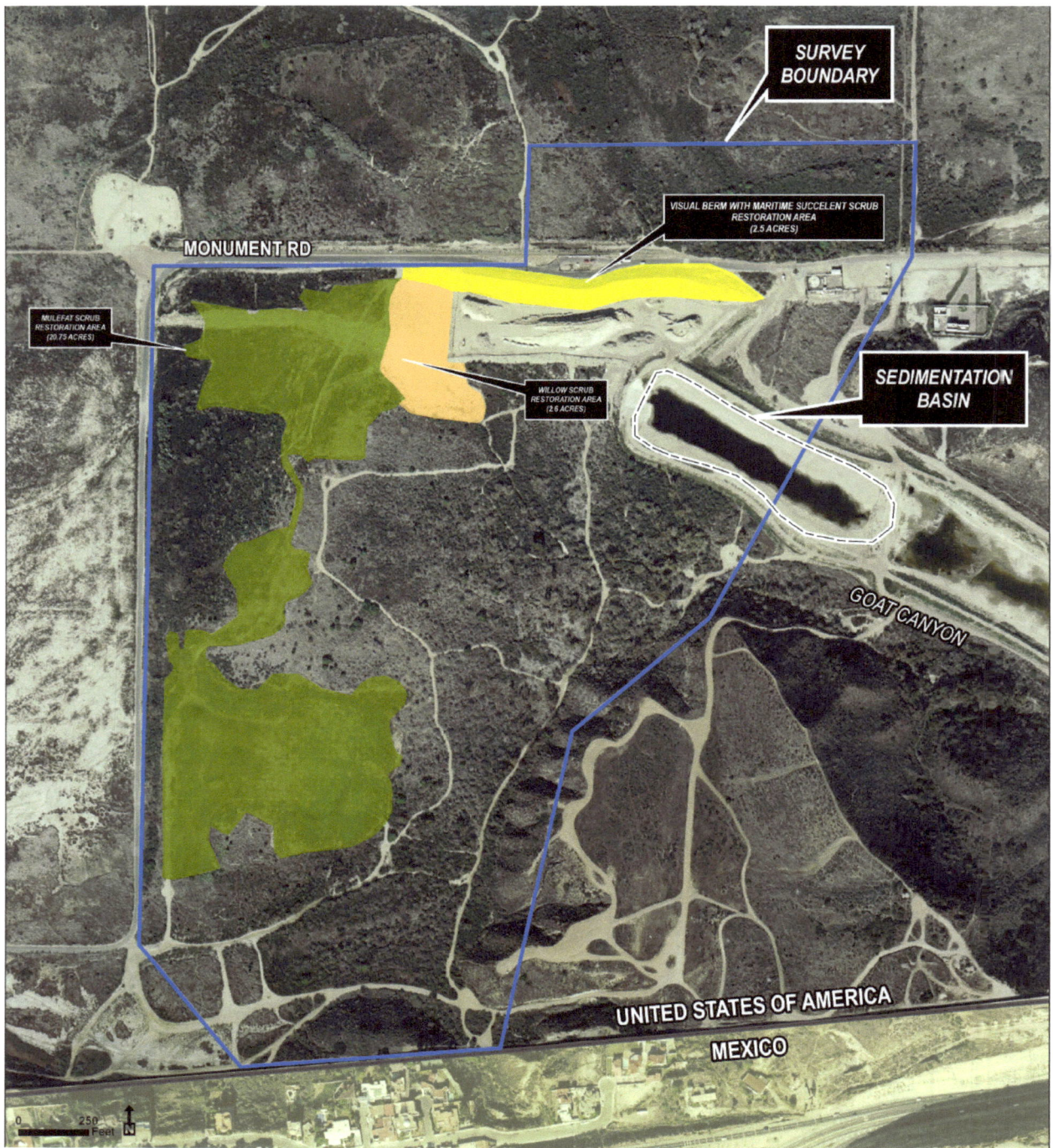

Figure 33. Goat Canyon Mitigation Area

LESSONS LEARNED

The mitigation area has suffered from the prolonged drought. Plans to use the clean water after passing through the sedimentation basins for supplemental irrigation have not succeeded and some parts of the mitigation site are declining. Future restoration plans in the region should not count on rainfall and runoff for habitat support.

Disposal of sediment captured each year is a challenge to California State Parks that manages the sedimentation basins. The sediment is silty and mixed with plastic trash and debris, some of which is too small to be screened. Consequently, it is of low quality as construction material. Disposal at a landfill can cost more than $1,000,000 annually. A one-time test disposal of the sediment in the nearshore marine habitat showed that the plume associated with this placement was well within the range of conditions associated with natural river flow events, with the coarser sediments being transported to the beach and finer sediments being transported via waves and currents offshore. That project, known as the Tijuana Estuary Fate and Transport Project, is presented later in this document.

Several other important lessons were learned during the planning, permitting and mitigation of the Goat Canyon Sediment Basins Project. Two issues occurred during planning and permitting that had direct impacts to the project:

1) delays in Section 7 Endangered Species Act consultations with the USFWS; and
2) the decision by the Coastal Commission to not permit the raising of Monument Road.

The Section 7 consultation with the USFWS was delayed over concerns that a small portion of one of the occupied least Bell's vireo territories would be impacted by the project. The project was delayed while the sediment basins were redesigned and another year of focused surveys for the vireo were conducted. Thus, the project was delayed approximately one year. During construction in 2004–2005 the partially constructed basins were overwhelmed by a large storm event and approximately 20 acres of high marsh and salt pan were buried by up to two feet of sediment.

The Coastal Commission decision to not elevate Monument Road has had direct impacts to the operation of Border Field State Park. The park still does not have year-round access and polluted water entering the estuary from Yogurt Canyon poses health issues as well as limits access. The decision was based on the interpretation that the Coastal Act does not consider transportation projects as essential public projects. The impacts on wetland habitats along the road were deemed too great to allow. Sediment deposition along Monument Road prior to construction of the sediment basins has altered these habitats, much of which is now disturbed upland.

While an event such as the high-rainfall year of 2004–2005 could not be foreseen, a more proactive approach by the design team, such as a project design with no impacts to endangered species, could have been adopted. Conversely, early buy-in from regulators about potential impacts (appropriately mitigated), might have avoided the loss of valuable wetland

habitat. The latter approach might be particularly relevant for projects associated with the NERR, given its formal role, along with the Coastal Commission and Coastal Conservancy, in the State of California's Coastal Zone Management Program.

15 FENTON QUARRY RESTORATION PROJECT

Project Engineers:
Group Delta Consultants
Prime Contractor:
Terra Movers
Landscape Contractors:
Natures Image/D&D Landscape Construction
Restoration Ecologist:
Chris Nordby, Tierra Environmental Services
Project Administration:
Mayda Winter, SWIA
Date: 2001–2006
Funding Agency:
California State Coastal Conservancy
Reported in:
Assessment of the Fenton Quarry Habitat Restoration Project. Tierra Environmental Services. March 14, 2006.

Fenton Quarry, located on the western slope of Spooner's Mesa above Goat Canyon, was operated as a sand and gravel quarry until it was abandoned in the 1980s. Although not originally envisioned as part of the Model Marsh project, it was integrated into it once the quarry was identified as a disposal site for sediment excavated during construction.

METHODS

The sediment that was excavated during construction of the Model Marsh was stockpiled at the mouth of Goat Canyon and the following year was transported into the quarry, compacted and planted with maritime succulent scrub species (Figure 34). Approximately 4.3 acres of maritime succulent scrub habitat was restored. Because the sediment from the Model Marsh was saline, a four-foot thick cap of native material was set aside until the marsh sediments were in place, then placed over the saline material to allow native upland species to grow. The site received supplemental irrigation during the first two years of establishment.

The intent of the quarry restoration project was to restore the topography of the site to pre-mining conditions and to establish high quality maritime succulent scrub habitat. As part of the site preparation, all existing scrub was brushed and stored separately within the stockpile area for later use as mulch for the restoration effort. Several overarching restoration goals were identified including:

- Restore areas of former coastal bluff affected by sand and gravel extraction activities;
- Restore areas of former maritime succulent scrub affected by sand and gravel extraction activities;
- Provide habitat for the state-listed threatened coastal California gnatcatcher (*Polioptila californica californica*) and other animals associated with maritime succulent scrub;
- To the extent possible, salvage, incorporate and reintroduce existing native plant individuals present on-site prior to project construction.

Approximately 100,000 yd³ of stockpiled soils were transported into the quarry using belly dump trucks, excavators, bulldozers and front loaders. Slopes were stabilized using layers of geotextile plastic. Once the slopes were deemed stable by the geotechnical consultant, non-saline soils were used to cap the saline soils from the Model Marsh. A uniform four-foot thick layer of capping soil, excavated from the quarry for this purpose, was distributed over the top and slopes of the restored quarry. Once in place, the capping soils were compacted and contoured.

FIGURE 34. FENTON QUARRY

The approximately 100,000 yd³ of soil did not completely fill the quarry as planned by the project engineer. The engineer had developed the plans based on the amount of soil excavated during construction of the Model Marsh—approximately 133,000 yd³. This discrepancy was never resolved.

Once the soils were contoured, the approximately 4.3-acre area was hydroseeded with maritime succulent scrub plant species and the stockpiled scrub mulch was spread over selected locations to facilitate establishment of desired species and to reintroduce microorganisms. The slopes of the maritime succulent scrub hydroseed mix

were covered with jute netting for erosion control. In addition, 1,050 individual golden club cactus (*Bergerocactus emoryi*) were planted on-site. These were planted from cuttings, each approximately 18 inches long, collected from naturally occurring patches in the project vicinity. Of the 1,050 individuals, 525 were treated at a nursery and allowed to develop roots and the remaining 525 were planted immediately after harvesting. Rooted cuttings were laid lengthwise with roots covered lightly by soil. Unrooted cuttings were planted vertically approximately six to eight inches deep. The site was irrigated with overhead sprinklers to ensure survival of cuttings and germination of the hydroseed mix. These activities were completed February 2002 which signified the beginning of the five-year plant establishment period. A subcontractor installed the initial restoration components, including planting, seeding and maintenance. However, in 2003, this subcontractor quit the project after a dispute with the prime contractor. They were replaced by another firm, who was responsible for maintenance, installation of additional irrigation lines, and replanting of selected species.

Remedial work was conducted during the plant establishment period in areas where initial revegetation efforts had not been successful. These efforts consisted of the addition of gypsum to reduce salts that had leached from the lower, saline soils, and hand-seeding open areas by first scarifying the soil surface with hand cultivators. Additional one-gallon container stock was planted as well as additional cuttings of golden club cactus. Regular monitoring site visits were conducted over the plant establishment period. Originally proposed for five years, project monitoring was terminated after three years as a result of achievement of project goals, including use of the restored site by coastal California gnatcatcher. Irrigation was discontinued after the second year of the project.

Monitoring included qualitative and quantitative methods. Qualitative monitoring included visual assessment of the overall condition of the site, and health and vigor of hydroseeded plants, cuttings and container stock. Photographs were taken from permanent stations to document canopy development. Quantitative monitoring included canopy development, mean percent cover, and percent cover by species along 11 transects using the point intercept method and cover class method, and direct count of planted golden club cactus.

RESULTS

Overall mean cover as determined by point intercept along 11 transects was approximately 52%. This was deemed to be within the characteristics of naturally occurring maritime succulent scrub vegetation which is open with cover varying from 25% to 75%. Dominant species included coastal sagebrush (*Artemisia californica*), California encelia (*Encelia californica*), California buckwheat (*Eriogonum fasciculatum*) and deerweed (*Lotus scoparius*).

Planted golden club cactus were censused six weeks after planting. Of the 525 rooted cuttings installed, 57 (10.9%) established successfully and the remaining 468 (89.1%) had died. Of the 525 unrooted cuttings installed, 424 (80.8%) had established successfully while the remaining 101 (19.2%) had died. All individuals planted as unrooted cuttings had at least one pup. Due to the high mortality of rooted cuttings, an additional 238 unrooted individuals in April 2003 were installed. Of those, 193 (81.1%) lived and 45 (18.9%) died. Thus, it is evident that planting unrooted cuttings immediately after harvesting is the preferred method for establishing this species.

Coastal California gnatcatcher was detected foraging within the restored quarry on several monitoring visits. After the first observation of this species, monitoring was confined to its non-breeding season.

LESSONS LEARNED

There were a number of lessons learned in the Fenton Quarry project and some mysteries that may have been unique to the project. The main lesson learned is that saline soils excavated during construction of estuarine restoration projects can be disposed of at upland sites and successfully revegetated. However, diligence is required. Despite a four-foot deep cap of non-saline upland soil, salts wicked into the root zone of some shrubs and addition of gypsum was required to neutralize those salts. Plants burned by high salt concentrations had to be replaced after gypsum addition.

Cactus plantings demonstrated that unrooted golden club cactus cuttings were far more successful than rooted cuttings. Whether this is species-specific or applicable to other cactus species may require additional experimentation.

The discrepancy over the quantity of soil excavated versus the amount available for restoration one year later was never resolved. Typically, soil that is excavated tends to swell and increase in volume rather than shrink and decrease in volume. Thus, the Model Marsh contractor either erred in their measurement of the quantity excavated or the soil shrunk during the year it was stockpiled or some person or persons removed some of the stockpile during that year.

The relationship between subcontractors and the prime contractor can always be contentious and is not unique to this project. A subcontractor quitting, however, is a rare occurrence and can delay project construction.

It is hopeful that this project can be applied to other restorations, such as the Tijuana Estuary–Friendship Marsh Restoration. A large, former quarry known as the Nelson Sloan Quarry exists in the Tijuana River Valley where soils excavated for the Friendship Marsh project could be disposed and revegetated similarly. A restoration plan for this quarry has been prepared as presented later in this section.

16 TIJUANA VALLEY INVASIVE PLANT CONTROL PROJECT

Lead Agencies:
> California State Parks, City of San Diego, County of San Diego

Project Consultants:
> Tierra Environmental Services, Boland Ecological Services

Restoration Ecologist:
> John Boland, Boland Ecological Services

Project Administration:
> Mayda Winter, SWIA

Date: 2002–present

Funding Agencies:
> California State Coastal Conservancy , U.S. Fish and Wildlife Service,
> Regional Water Quality Control Board, City of San Diego

Reported in:
> Various Reports—Boland Ecological Services

The Tijuana Valley Invasive Plant Control Project (Invasives Project) was a multi-year effort by the Coastal Conservancy, the USFWS, the RWQCB, the City of San Diego, and SWIA to control three primary invasive species that are prevalent in the river valley. The three species are giant reed (*Arundo donax*), castor bean (*Ricinus communis*) and salt cedar (*Tamarix ramosissima*). These species degrade the habitats they invade by displacing native vegetation, lowering insect food supply for birds, reducing groundwater, and increasing flood and fire hazards. The objective of the program is to enhance the native habitats of the valley including willow woodlands, mule-fat scrublands and salt marshes, by removing these damaging exotic species. The Invasives Project was conducted throughout the river valley in multiple phases.

METHODS AND RESULTS

Phase 1 Funded by the Coastal Conservancy and USFWS 2002. The program was initiated in 2002 with funding from the Coastal Conservancy and the USFWS administered by SWIA. John Boland, under contract to Tierra Environmental Services, mapped the distribution of the target species in the valley and a control plan was developed in consultation with a Technical Advisory Group, which included members from the resource agencies, land owning agencies and experts in the field of exotic plant control. The plan described the methods to be used to control the target invasive species. The program received support from land owners and managers in the valley and by 2003 had obtained all the necessary environmental permits: NEPA—Categorical Exclusion (USFWS); CEQA—Negative Declaration (Coastal Conservancy); and Streambed Alteration Agreement (CDFW).

Initial work focused on several research questions including:

- What is the best method to control salt cedar in saline situations?
- How successful are control efforts conducted during winter?
- Can the control still be effective by leaving the giant reed biomass on site and thereby reducing costs?
- How should a treatment site be prepared to allow the site to be restored?

Phase 2 Funded by the RWQCB 2004–2007. The RWQCB funded exotic control efforts for three years under the Proposition 13 Watershed Protection Grant Program that included control of the three target species and active restoration, including planting treatment areas with native riparian species and monitoring their success. The focus of this phase of work was to quantify the following:

- Acreage of estuarine and riparian habitat that were treated;
- Volume of invasive biomass that was removed;
- Number of native plants installed;
- Length of river treated;

- Effectiveness of the herbicide treatment;
- Growth rate and survivorship rate of planted native;
- Natural recruitment of native plants following treatment.

The methods used to control the target plant were specific to each species and included direct foliar spray (castor bean), manual cutting of stems followed by immediate application of herbicide to freshly cut stumps (salt cedar and giant reed), or manual cutting of stems with later application of foliar spray to the new growth (giant reed). The County of San Diego granted permission to treat giant reed and castor bean on their property but would not allow treatment of salt cedar.

Phase 3 Funded by RWQCB 2006–2008. The RWQCB funded exotic control efforts for an additional two years under a Proposition 40 Non-point Source Pollution Control grant. The goals of the project were:

- To carry out initial treatment of 434 acres of riparian and estuarine habitats;
- To retreat 874 acres treated in Phase 2;
- To plant additional native plants and monitor their survival, and;
- To plant 9.5 acres of riparian habitat with 11,066 willow and mule-fat cuttings.

Giant reed and castor bean were treated using the foliar spray and leave-in-place method. Salt cedar was treated using the cut-stump and apply herbicide method. The project exceeded its goal of 434 acres by treating 590 acres.

Phase 4 funded by SANDAG TransNet Environmental Mitigation Program 2009–2010. During Phase 4, 87.5 acres of riparian habitat were treated and an additional 1.5 acres were planted with 1,200 willow and mule-fat cuttings.

Phase 5 Funded by the USFWS San Diego Wildlife Refuge Complex 2010–2012. Phase 5 consisted of three sub-phases with treatment funded annually for each year. During the 2010–2011 treatment year, a 10-acre site in the Tijuana Slough National Wildlife Refuge was treated to control giant reed, Brazilian pepper tree (*Schinus terebenthifolius*), sea fig (*Carpobrotus* spp.) and myoporum (*Myoporum laetum*). All species were reduced to zero percent cover.

During the 2011–2012 treatment year a 21-acre site located in the eastern portion of the Refuge was retreated for giant reed and castor bean. This site had been treated previously in Phase 3. Treatments were largely successful; however giant reed resprouted an average of 27% and follow-up treatment was recommended.

Between 2012 and 2013, a 2.5-acre parcel located in the eastern portion of the Refuge was treated for salt cedar, giant reed, Brazilian pepper tree, myoporum, castor bean and sea fig. All were successfully treated. There was no revegetation conducted in Phase 5.

Tijuana River Valley Maintenance Project 2013–2018. In addition to the five phases of the Invasives Project, the City of San Diego Stormwater Division was required by the permitting agencies to mitigate for impacts associated with clearing Smuggler's Gulch and a portion of the Tijuana River referred to as the "Pilot Channel." Both in-channel and out-of-channel mitigation was required. Boland Ecological Services and SWIA were funded to complete the out-of-channel mitigation. The out-of-channel mitigation requires that giant reed, castor bean and salt cedar be treated in a 4.31-acre site. The City is required to conduct follow up treatment for five years. An interim report for the first seven months of the project documented that 4.96 acres were treated using the foliar spray and leave-in-place method after which the contractors returned, cut and chipped dead biomass and transported off-site.

LESSONS LEARNED

The effectiveness of two herbicides— Imazapyr (i.e., Stalker) and Glyphosate (i.e., Aquamaster)—was analyzed for control of salt cedar. Both were equally effective, killing approximately 90% of the plants treated without resprouting.

The most effective time of the year to control giant reed is in the fall from approximately mid-September to the end of October. During this period, plants that are cut and treated with herbicide grew back quickly and could be controlled by foliar spray. Plants treated similarly in the winter did not resprout quickly and success could not be ascertained until spring making a prolonged, two-phase treatment necessary. It was concluded that cutting, treating and removal of above-ground biomass was the most effective method for control. Although foliar spraying of giant reed was the least expensive control method, initial efforts concluded that it was also the least effective. In later phases Dr. Boland revised his conclusions regarding foliar spray and biomass removal of giant reed. Glyphosate herbicide was sprayed on stands of giant reed and the biomass was left in place. This proved to be effective and the least expensive as biomass removal was an expensive undertaking. However, retreatment of giant reed was essential to successful control. Foliar application of Glyphosate was also effective in killing castor bean. However, the seed bank for this species remains viable and germination requires that follow-up spraying be conducted. Although the majority of the plants are killed with one or two applications, giant reed and castor bean populations require retreatment every one or two years. Salt cedar and other tree species, such as Brazilian pepper, were stump cut and painted with Imazapyr. Treated areas of this species also need to be revisited to catch resprouts. Dr. Boland recommends that all sites be revegetated with natives to compete with invasives that attempt to reestablish.

17 TIJUANA RIVER VALLEY REGIONAL PARK TRAILS AND HABITAT ENHANCEMENT PROJECT

Project Consultants:
  Greystone Environmental Consultants, Kimley-Horn and Associates,
  Rick Engineering Company, WRC, Van Dyke LLP, Tierra Environmental Services
Restoration Ecologist:
  Chris Nordby, Tierra Environmental Services
Project Client:
  County of San Diego, Department of Parks and Recreation
Date: 2004–2010
Funding Agency:
  California State Coastal Conservancy
Reported in:
  Recirculated Draft EIR. Tijuana River Valley Regional Park Trails and Habitat
  Enhancement Project. Kimley-Horn and Associates. August 2006.

The Tijuana River Valley Regional Park Trails and Habitat Enhancement Project (Tijuana Trails Project) was a multi-faceted project undertaken to improve the Tijuana River Valley Regional Park (TRVRP). There were three primary project components:

1) designation of a formal trails system within the TRVRP;

2) Restoration of approximately 60 acres of riparian habitat west of Dairy Mart Ponds; and

3) feasibility assessment of construction of athletic fields on agricultural fields east of Hollister Street and north of Sunset Avenue, all located within the TRVRP boundaries.

The project analyzed several alternatives in the project EIR, including the No Action Alternative and the Proposed Project Alternative. Only the Proposed Project Alternative is presented here for the sake of brevity.

METHODS AND RESULTS

The over-arching objective of the project was to implement a trails and habitat restoration effort in the TRVRP. The project was intended to create, enhance and restore natural habitats within TRVRP while optimizing the recreational use of the park and ongoing border protection activities. This was to be achieved through creation of a formal trail network and revegetation of numerous unauthorized trails and dirt pathways. In addition, the project was planned to provide public access to the coast and linkages to regional trail systems, such as the California Trail and Bayshore Bikeway and planned linkages to the east of TRVRP. The project focus was on the establishment of the formal trail system and did not address potential restoration opportunities, such as restoration of Spooner's Mesa and removal

of existing berms in the valley for flood control. The project included planning-level assessment of the potential for future recreational opportunities, such as athletic fields; however, it was acknowledged that funding for such amenities was not available from the County at the time of the study.

The project consisted of the following components:

Development of Formal Trail System. At the time the project was initiated, there were approximately 71.5 miles of unauthorized trails and dirt roads in the TRVRP. These impromptu trails had been created by hikers, equestrians, U.S. Customs and Border Protection Services (Border Patrol), illegal border crossings and other users. Use of these informal trails resulted in degradation of sensitive habitats and impacts to sensitive plant and animal species.

The proposed project would create a 22.5 miles of formal trails (thus eliminating 49 miles of informal trails). Within these formal trails, users would be designated. For example, some segments would be multi-purpose (pedestrians, cyclists and equestrians); some would be multi-purpose but shared with the Border patrol; and some would be restricted to equestrians and pedestrians.

Revegetation of Informal Trails. Of the 49 miles of trails closed as a result of the designated formal trail network, approximately 34 miles of trails located in riparian and upland habitats would be either passively or actively restored. The remainder would be retained exclusively for use by the Border Patrol.

Wetland Restoration. Restoration plans encompass approximately 60 acres of habitat west of Dairy Mart Road, including freshwater marsh, riparian and coastal sage scrub habitats. Detailed restoration plans were developed by Tierra Environmental Services (Chris Nordby) in association with Rick Engineering (hydrology), WRC (hydrology) and Van Dyke LLC (landscape architects). The plans included planting palettes, spacing and densities for southern willow scrub, southern cottonwood–willow riparian forest, freshwater marsh, mule-fat scrub and Diegan coastal sage scrub habitats; irrigation design and schedule; and a long-term monitoring and maintenance plan.

Development of Staging Facilities. Staging areas included establishment of a two-acre eastern trailhead staging area along the west side of Dairy Mart Road north of the Tijuana River.

Construction of Bridge Crossing. Bridge construction would consist of a steel, semi-truss bridge multi-use recreation bridge over the Tijuana River approximately one mile west of Hollister Street. This bridge was to be constructed following dredging of the Tijuana River Pilot channel conducted by the City of San Diego.

Development of Recreational Opportunities. This included establishment of interpretive and directional signage, benches, bird observation blinds, scenic vistas and outlooks.

An EIR was completed by Greystone Environmental Consultants in 2006, and subsequently revised and redistributed by the County with revisions done by Kimley-Horn and Associates. The EIR identified project impacts and mitigation for a number of resource areas.

Based on the EIR and technical appendices, the formal trails system was established and informal trails were closed. Informal trails were mostly left to revegetate passively. The eastern staging area was designated.

The 60-acre restoration west of Dairy Mart Road, the multi-use bridge over the Tijuana River, and the athletic fields were not constructed and have not been constructed to date.

LESSONS LEARNED

Although the project objectives might not appear to be controversial, those familiar with the river valley recognize that any suggested change in land uses can generate resistance from various stakeholders.

The project EIR identified several areas of controversy that typify habitat restoration/alteration proposals in the Tijuana River Valley. Equestrians noted their long-standing presence in the valley and objected to closure of many of the informal trails, insisting that the majority of the trials be retained. They felt that many of the trails should be for

equestrian use only as they were concerned that horses would be negatively impacted by other users, especially cyclists. Equestrians asserted that horses provided benefits to the local community and to public health and safety and contested the assertion that horses have a negative effect on the environment of the valley. This opinion was in direct conflict with environmentalist groups who wanted to close certain popular equestrian trails for resource protection, especially protection of least Bell's vireo, from impacts related to human use, trail maintenance and invasive species. Public meetings for the project were contentious.

The U.S. Customs and Border Patrol (CBP) objected to the closure of many of the informal trails that they used, citing their potential inability to protect the area. CBP issues were partially resolved by designating some segments as CBP only and others as shared multi-user/CBP trails.

Funds for implementing restoration of closed trails were insufficient and most were left to revegetate passively. This has led to highly variable results with some closed trails unvegetated, some supporting mostly weed species, and others supporting monotypic or nearly monotypic native species, such as coastal goldenbush (*Isocoma menziesii*). The 60-acre riparian restoration was not funded and the athletic fields were deferred to later study.

Although the creation of the Tijuana River Valley Recovery Team (TRVRT) has provided a forum for stakeholders and thereby facilitated a more cooperative spirit among valley residents, disagreement in management strategies continues. The Tijuana Estuary is one of the few places in California that allow horses on the beach and equestrians are concerned that TRNERR, CSP and the USFWS will implement management plans that impact that use based on public recreational uses and impacts to threatened and endangered species. Implementation of other land use plans for the valley, including the County of San Diego Tijuana River Valley Master Plan, should anticipate similar public debate and plan accordingly.

The CBP has always had a high profile in the valley. Although border crossings by undocumented migrants have decreased relative to peaks in the 1980s and 1990s, the BIS project demonstrates congressional support for increased CBP presence. With the new fences and high-speed roadways, the creation of the TRVTF and the adoption of the Tijuana Valley Trails Project it is hoped that the CBP will no longer use unofficial roads in conducting their patrols of the river valley. However, the recent vehicular activity by CBP into the restored Fenton Quarry illustrates the ongoing challenges with maintaining official roads in the valley.

18 TIJUANA ESTUARY FATE AND TRANSPORT PROJECT

Lead Agencies:
California State Parks, California State Coastal Conservancy,
U.S. Geological Survey (USGS)

Project Consultants:
Moffatt & Nichol, Nordby Biological Consulting, Nautilus Environmental,
Boland Ecological Services, Avian Research Associates, Deltares

Others:
Scripps Institution of Oceanography

Project Administration:
Mayda Winter, SWIA

Date: 2008–2017

Funding Agencies:
California State Coastal Conservancy, USGS

Reported in:
Warrick, J.A. 2013. Dispersal of Fine Sediment in Nearshore Coastal Waters. Journal of Coastal Research, V. 29, No.3. pp579-596. May 2013.

Warrick, J. A., K. Rosenberger, A. Lam, J. Ferreira, I. M. Miller, M. Rippy, J. Svejkovsky, and N. Mustain. 2012. Observations of Coastal Sediment Dynamics of the Tijuana Estuary Fine Sediment Fate and Transport Demonstration Project, Imperial Beach, California. Published by U.S. Geological Survey (USGS). USGS Open-File Report 2012-1083. 29p.

Tijuana Estuary Sediment Fate & Transport Project Biological Resources Monitoring Report. Nautilus Environmental. September 2013.

Deltares. Tijuana River Coastal Suspended Sediments—Executive Summary. November 2015.

Everest International Consultants, Inc. Tijuana River Estuary Fate and Transport Study. Summary Report for Coastal Managers. March 2017.

The Tijuana Estuary Fate and Transport Project (Fate and Transport Project) was a cooperative effort of the California State Parks, in partnership with the State Coastal Conservancy the California Coastal Sediment Management Workgroup (CSMW), the USGS, the TRNERR, and SWIA. The project addressed the potential for sediment, such as that captured by the Goat Canyon sediment basins or sediment excavated for wetland restoration, to be disposed of in the nearshore environment where it would nourish the beach. The premise for such disposal is the documented ability of nearshore marine organisms to adapt to natural sediment deposition in the nearshore environment. The project sought to demonstrate that placement of sediment that does not currently meet regulatory criteria for nearshore placement would not result in impacts to the marine environment and that such disposal should be reconsidered.

Currently, the U.S. Army Corps of Engineers (USACE) precautionary rule of thumb requires sediment used for beach nourishment to be comprised of less than 20% fines (silt and clay). Sediment containing higher than 20% fines is believed to result in environmental degradation. This so-called 80/20 (coarse-to-fine sediment) guideline was originally used as a threshold due to concern that pollutants associated with the fines fraction might be present in sufficient quantity to produce toxicity. However, this rule is now applied also to uncontaminated sediment, in part because fine-grained sediment may lead to turbidity and other ecological impacts uniquely associated with fines.

Unfortunately, much of the sediment available for beach nourishment does not meet the 80/20 grain size standard and must be transported inland to landfills at high cost. This project was conducted in order to evaluate whether the current guidelines are appropriately protective or overly conservative for California coastlines. Since the project was designed as a sediment fate and transport study, it afforded an opportunity to investigate how fine sediment moves in a coastal setting, and, thus, provided valuable information for future sediment management activities.

The project utilized sediment obtained from the Goat Canyon sediment basins. Sediment was tested for priority pollutant contaminants and grain size characteristics, and was approved for aquatic disposal by the USACE. Subsequently, sediment was sorted at an existing staging area to achieve the desired ratio of fines to coarse sediment (up to 50% fines). Sorted sediment was then transported from the staging area near the Goat Canyon retention basins to the beach approximately 0.5 mile south of the Tijuana River mouth via a haul truck route along a dirt road that serves as a horse trail (Horse Trail Road) and deposited on the beach (Figure 35).

The original scope of the project included three phases within one wet season, and a deposition of a total volume of 60,000 yd^3 of sediment. The project plan was altered

FIGURE 35. PLACEMENT OF SEDIMENT ON THE BEACH

during the project due to logistical and financial constraints. The initial phase of the altered project scope included approximately 10,000 yd³ of sediment that was deposited from November 3 to November 13, 2008 (Phase I). After completion of Phase I in 2008, a state funding freeze occurred and the project was halted. Funding from alternate sources in 2009 allowed the project to be restarted. The original scope was scaled back to include one additional phase. Phase II occurred from September 21 to October 2, 2009 in which approximately 35,000 yd³ of sediment was placed at the same placement site. Placement of the material in the intertidal occurred during periods of mid-range and low tides at both events. A total volume of roughly 45,000 yd³ was utilized in the demonstration project.

METHODS

Dispersion of the materials placed at the site was monitored in the nearshore environment by the USGS according to the Physical Science Monitoring Plan. This monitoring was conducted to investigate transport dynamics of fine-grained sediment along the

shoreline and address the following specific questions about fine-grained sediment in California nearshore waters:

- What are the transport pathways and fate of fine-grained sediment introduced at the coastal shoreline?
- How do environmental and project variables, such as sediment placement volume, percent fines, waves, currents, and shelf setting, influence the rates and modes of transport and eventual fate?

USGS monitoring included seafloor mapping and turbidity plume tracking. Plume tracking data was obtained using nearshore moored instrumentation, vessel surveys, surf-zone sampling, aerial survey data, and other locally available data (including data from Southern California Coastal Ocean Observation System and National Oceanic and Atmospheric Administration program sources). Results of the Physical Monitoring Science Program are presented in Warrick 2013.

In addition to the physical aspects of the project, placement of fine-grained sediments on California beaches also has the potential to impact the existing biota. A biological monitoring plan was developed and implemented that focused on two general habitats of interest: the intertidal zone in which the sediments would be placed and the nearshore subtidal areas that are potentially impacted by sediments as they are transported offshore or alongshore by natural processes. Impacts to these habitats also have the potential to impact

the broader ecological system, in particular, that of shorebirds that feed upon intertidal invertebrates. The biological monitoring plan sought to answer the following questions:

- Is the benthic macroinvertebrate fauna of the beach impacted in the long-term by fine-grained sediment placement?
- If so, what are the spatial and temporal dimensions of the impact?
- Are offshore sand dollar (*Dendraster excentricus*) beds impacted by the sediment placement, either in terms of the spatial band they inhabit or in terms of population attributes?
- Are bird foraging patterns influenced by the project?

Several types of monitoring were conducted for the project. The USGS monitored the transport and fate of the sediment using four primary methods:

1) surf-zone sampling of swash and beach sediment;
2) airborne remote sampling of coastal water color and temperature;
3) nearshore water and sediment sampling from a small boat; and
4) oceanographic instrumentation on moored buoys and benthic tripods.

Scripps Institution of Oceanography conducted monitoring of bacteria levels in the surf zone. Nordby Biological Consulting conducted monitoring of the sediment placement on the beach and the resulting plume as a condition of the RWQCB permit for the project. Boland Ecological Services monitored the effect of the project on shorebirds before, during and after the project. Avian Research Associates monitored the effect of trucks accessing the surf zone on the threatened western snowy plover and Nautilus Environmental monitored the effect of the project on beach profiles and intertidal invertebrates.

The final component was a modeling exercise led by Deltares, where a plume from a beach placement was compared to a plume from a river flow event. Modeled plumes were compared under a wide variety of ocean conditions (e.g. wave height, swell direction, wind, and temperature).

RESULTS

Fate of Sediment. Swash sampling and ocean color remote sensing results suggest that the initial transport of fine sediment was strongly influenced by longshore currents of the surf-zone that were established by the approach angles of waves. Measurable fine sediment concentrations persisted in the surf-zone for several days after nourishment due to the slow winnowing of fine sediment that remained within the beach foreshore (Figure 36). Immediately offshore of the surf-zone, fine sediment was pulled offshore by rip currents and settled on the seafloor. Settlement on the seafloor was greatest immediately offshore of the nourishment site, although a mass balance of sediment suggested that the majority of fine sediment moved far away (more than two km) from the nourishment site to depths greater than

10 m where similar fine sediments make up a substantial portion of the seabed. Thus, the fine sediments deposited in the surf-zone were transported to areas of naturally occurring fine sediment where they settled. A model of sediment dynamics was constructed by USGS for use in further research.

Project Effects on Shorebirds. In order to assess the effects of sediment deposition on shorebirds, shorebird populations were monitored at the 600-meter long deposition site and seven 600-meter long control sites before, during and after sediment placement in both 2008 and 2009. One control site was located south of the deposition site and six were located north of the deposition site. The results were analyzed using the before-after-control-impact (BACI) experimental design.

Shorebirds were monitored weekly from September 3 to December 3, 2008 and from September 7 to November 16, 2009. Each site was surveyed before, during and after sediment placement on the beach. All surveys were conducted approximately 1.5 hours after high tide for approximately two hours.

ANOVA tests of the number of foraging shorebirds at site showed no significant differences between before and after abundances at the impact site when compared to the control sites. These results indicated that sediment deposition and movement did not have long-term impacts to foraging shorebirds.

FIGURE 36. IMAGERY OF PLUME

Project Effects on Benthic Macrofauna. Like the shorebird monitoring program, benthic macrofaunal monitoring employed the BACI experimental method. Before populations were monitored in July and October of 2008 prior to Phase I sediment placement to establish baseline conditions. Additional monitoring events occurred in May, July, and September of 2009 following Phase I placement. Phase II sediment was placed in September and October of 2009, and macroinvertebrate monitoring events were conducted post-placement in May, July, and October of 2010.

Benthic invertebrates were collected from four sites: the placement site, a control site south of the placement site, a control site north of the placement site, and a third control location at Silver Strand State Beach. Invertebrates were collected at randomized sampling locations along sampling transects that spanned the upper beach to the shallow intertidal habitat using a six-inch diameter core expressed into the sediment to a depth of 20 centimeters. Individuals retained on a one-millimeter mesh screen were fixed in the field and counted and identified in the laboratory. Abundance and wet weight biomass were measured for each major taxonomic class. Sand dollars and Pismo clams were qualitatively assessed along transects in the intertidal zone.

Abundance and biomass data were sorted by season and analyzed using two-way ANOVA tests. Although the analyses were complicated and contained considerable variability, there was no significant difference in abundance and biomass between the deposition site and control sites, indicating that long-term effects of sediment deposition on benthic macroinfauna were not detected.

Project Effects on Western Snowy Plover. Western snowy plovers (*Charadrius alexandrinus nivosus*) were monitored from March through October of 2008 prior to project initiation. During placement in 2008, surveys were employed for the first two days of sediment deposition on the beach.

During each monitoring event, a 600-meter (1,956-foot) stretch of beach (the project area) was surveyed and western snowy plover abundance, distribution, and any behavioral reactions to the construction activities were noted. The number of individuals and their behavior were consistent with pre-project observations of normal foraging and roosting activities at the placement location. As a result of the apparent lack of impact, daily western snowy plover monitoring was discontinued.

Comparisons of Modeled Nourishment and River Plumes. The modeling exercise demonstrated that, except for the immediate vicinity of the sediment deposition on the beach, sediment concentrations in the nourishment plume were well within the range of those observed within the plume that would develop with discharge of the Tijuana River after a storm.

LESSONS LEARNED

The most obvious lesson learned from the Fate and Transport Project is that the placement of up to 45,000 yd³ of sediment composed of approximately 50% fine-grained material had no discernable negative effects on the physical and biological receptors in the shore and nearshore environments. Moreover, sediment placement activities such as this can be designed in a way that would resemble natural river discharge events, suggesting that biota in the project area would be adapted to (and likely need) inputs of fine sediments. Also, the coarse component (sand) of placed sediment can serve as beach nourishment and can be considered a beneficial reuse of excavated sediments. Despite these important results, it is likely that additional effort will be needed to convince regulators that rules governing disposal of such sediments should be relaxed in some cases, including those where few sensitive resources are located in the sediment placement vicinity. Discussions with regulators continue, and are expected to be important elements of TETRP.

Another lesson learned from the Fate and Transport Project involves constituents other than sediment in the Goat Canyon sediment stockpile. These sediments were sorted to remove trash and cobble, tested for grain size, bacteria and contaminants before they were disposed of in the nearshore. However, sorting failed to remove very small fragments of plastic that had broken off of the ubiquitous plastic debris retained by the sediment basins. These small but numerous fragments were observed along the shoreline after deposition events and were assumed to also occur in the nearshore surf wash. Recent literature has identified the negative impacts of plastic fragments on nearshore fish that ingest such fragments assuming that they are food. Clearly, wetland restoration projects with proposed nearshore disposal have a responsibility to avoid such impacts.

Although the sediment was tested for bacteria prior to deposition, monitoring by Scripps surprisingly indicated that concentrations of fecal indicator bacteria, in particular *Enterococcus*, briefly exceeded single-sample public health standards. The concentrations of *Enterococcus* were highly correlated to the deposition of fine sediment. Thus, public safety must be addressed in any future tests of the effects of such deposition.

The Fate and Transport Project required regional regulatory agency permitting and approval. Not all of these agencies were eager to conduct such a demonstration project with the fear that it would promote ocean disposal of fine-grained sediments. This reluctance complicated the permitting phase of the project as well as its implementation.

19 WRT TRANSITION ZONE RESTORATION

Principal Inverstigator:

Dr. Theresa Talley, California Sea Grant Extension Program, Scripps Institution of Oceanography

Date: 2009–2011

Funding Agencies:

NOAA, University of California Sea Grant

Reported in:

T.S. Sinicrope Talley, K.C. Nguyen, D.M. Talley, E. Ruiz, P.K. Dayton. Native plant diversity and introduced litter influence on the development of an urban coastal scrub ecosystem. Unpublished manuscript.

T.S. Talley. Year 3 Results & Management Implications for the Sea Grant Funded Project: Making restoration more predictable: Testing the contributions of planting diversity & introduced plant legacy efforts to recovering coastal ecosystems.

This study tested the hypothesis that the addition of organic litter and higher planting diversity would lead to faster development of most aspects of the ecosystem (soils, communities of plants and animals) compared to no litter addition and planting a monoculture. The restoration goal was to convert an area recently cleared of dominant invasive annuals to an ecosystem that resembles remnant native patches of coastal sage scrub and high marsh transition in the north arm of Tijuana Estuary. Ecosystem development using a comprehensive list of physical, soil, plant and faunal variables that reflect ecosystem processes were compared between the experimental and reference sites.

The experimental site was a one-acre parcel which had previously been a road waste dump site that was filled and, prior to the project, had been dominated by non-native garland chrysanthemum (*Glebionis coronaria*), slender-leaf iceplant (*Mesembryanthemum nodiflorum*), and crystalline iceplant (*M. crystallanum*). These and other weed species were cleared by hand at the beginning of the study (Figure 37). The soils contained chunks of asphalt, concrete, and metal pipe; and an invasives-dominated seedbank (as evidenced by regrowth of annuals outside of the experimental area).

The reference site was an adjacent area of remnant but disturbed coastal sage scrub and high salt marsh transition ecotypes, dominated by perennial natives. The reference site had been bisected by a paved

road that was later demolished and therefore contained some asphalt debris and disturbed patches similar to the experimental site. Despite this disturbance, the reference site supported a diversity of wildlife and native plants making the reference site an ideal restoration model for the experimental site.

The project was conducted at two elevations at the north end of Tijuana Estuary: coastal sage scrub (high elevation) and high salt marsh-upland transition (low elevation). Both high and low elevation habitats were included in the restored area and reference site.

METHODS

Ten replicate blocks containing treatment plots (80-centimeter diameter) were established in January 2009 in the experimental site at each of the two elevations. Experimental treatments were combinations of a litter treatment (plant mulch was added or not added) and a planting diversity treatment (zero, one, three or six species). Plants used in the planting diversity treatments were chosen based on common occurrence in the reference site and surrounding area. These included common coastal sage scrub species such as coastal sagebrush (*Artemisia californica*) and California buckwheat (*Eriogonum fasciculatum*) as well as common transition zone species, such as saltgrass (*Distichlis spicata*).

The experimental area was fenced to exclude herbivores and plots were watered as needed (at least two to three times per week during summer and fall, once per week during the rest of the year when no rain fell). Reference site plots consisted of 18 replicate plots (one m²) per elevation that captured the variability in plant diversity and composition used in the experimental site.

The study measured a number of physical parameters, soil and litter properties, plant properties and invertebrates in the plots for three years (2009–2011). Physical parameters included substrate temperature, humidity and light attenuation. Soil and litter properties included porewater salinity, soil moisture, soil organic matter content, texture, and concentrations of nitrate and ammonium. Plant properties included species composition and biomass and invertebrate sampling included pitfall trapping of ground-dwelling invertebrates.

RESULTS AND LESSONS LEARNED

The authors concluded that addition of litter played an important role in the early development of favorable soils (i.e., initiation of the soil microbial community, enhancement of fertility, and stress amelioration), in encouraging the growth of native plantings, in discouraging plant invasions, and in development of the invertebrate community, especially during the first year of the site when plant cover (i.e., productivity) was relatively sparse. By the end of the second year of ecosystem development, higher plant diversity, often higher plant productivity, more shade and

FIGURE 37. HAND CLEARING THE WRT SITE

humidity, cooler temperatures and lower salinity had more significant contributions to soil and invertebrate community development. After two to three years of ecosystem development, there were still differences between the created and reference site but the magnitude of differences was decreasing, especially in areas with litter and/or high plant cover.

The invertebrate communities of all areas (reference and experimental) were dominated by detritivores suggesting that these systems are detritally-driven. The addition of litter in the experimental site was associated with earlier appearances of predators (mostly spiders) than in plots without litter illustrating that litter may encourage the development of food webs due to added primary productivity, habitat complexity and/or mitigation of physical stresses. Conversely, litter was often associated with the presence and abundance of invasive invertebrates, such as the Argentine ant and the European earwig.

TIJUANA RIVER ESTUARY WATER QUALITY IMPROVEMENT AND COMMUNITY OUTREACH PROJECT

20

Lead Agency:
> TRNERR/California State Parks

Restoration Ecologists:
> Chris Peregrin, Greg Abbott, Cara Stafford

Project Administration:
> Debi Carey, SWIA

Date: 2009–2014

Funding Agency:
> U.S. Environmental Protection Agency, West Coast Estuaries Initiative

Reported in:
> Unpublished Summary of Project by C. Peregrin, CSP Ecologist.

The Tijuana River Estuary Water Quality Improvement and Community Outreach Project sought to further develop the TRNERR programs aimed at restoring and protecting the water quality, habitat and environment of the Tijuana River watershed. Efforts were also pursued to expand the ability of regional agencies to manage sediment in coastal ecosystems. A total of 106 watershed improvement projects were completed in the U.S. and Mexico to reduce erosion, address sediment, remove invasive species, establish native plants, repurpose waste tires, and remove trash. These projects relied on the efforts of over 3,000 volunteers.

METHODS AND RESULTS

The project entailed the restoration of five acres of mule-fat scrub upland and transition zone habitat in BFSP. A five-acre parcel was selected based on its high level of degradation and its proximity to sensitive least Bell's vireo habitat and nearby mitigation sites. This presented an opportunity to join and expand fragments of intact mule-fat scrub, southern willow scrub, and coastal sage scrub plant communities. Previous deposition of sediment and non-native, invasive species seeds left several acres dominated by invasive annual mustards (*Brassica* spp.) and nettle, (*Urtica* spp.) in addition to invasive woody perennials such as tree tobacco (*Nicotiana glauca*), giant reed, salt cedar, castor bean, and perennial pepperweed (*Lepidium latifolium*). Over the course of four years, approximately 9,200 native plants (cuttings and container plants) were planted to replace invasive weeds or to revegetate closed roads and trails. Challenges to plant establishment included foot and vehicle traffic, herbivory, sedimentation, high density invasive weed growth, and access to water. Despite higher mortality associated with these limiting factors, native plant cover surveyed over 12 permanent transects more than doubled from 2010 to 2014. A diverse assemblage of coastal sage scrub and riparian native perennials has replaced monotypic stands of nonnative annuals.

LESSONS LEARNED

The primary lesson learned from this project is that volunteer-based restoration efforts can be successful in the Tijuana River Valley and elsewhere. Although small in scale, these restoration projects result in incremental increases in the quantity and quality of natural communities. The Tijuana River Estuary Water Quality Improvement and Community Outreach Project also demonstrated the relationship between sediment and invasive, non-native plant species and techniques for their control.

21 NELSON SLOAN QUARRY RESTORATION PLAN

Lead Agencies:

> City of San Diego, Transportation and Storm Water Department—
> Environmental Lead

> County of San Diego, Department of Parks and Recreation—
> Long-term Management

Restoration Plan:

> Dudek

Dates: 2011–present

Reported in:

> Draft Conceptual Mitigation and Mine Reclamation Plan for the Nelson Sloan Quarry, Tijuana River Valley. Dudek. July 2011.

The Nelson Sloan Quarry is located southwest of the intersection of Dairy Mart Road and Monument Road in the Tijuana River Valley. It is part of the Tijuana River Valley Regional Park managed by the County of San Diego, Department of Parks and Recreation (County Parks). Like the Fenton Quarry, the Nelson Sloan Quarry was never reclaimed under the State Mine and Reclamation Act (SMARA). This restoration plan was prepared to comply with SMARA and to develop a plan whereby sediment excavated from maintenance of basins and channels in the river valley could be disposed of in a relatively inexpensive manner. The site was acquired by the State Coastal Conservancy which entered into an agreement with the County of San Diego that the property "be used for the purpose of habitat protection and open space." The City of San Diego is the environmental lead for the project.

The Nelson Sloan Quarry restoration site consists of approximately 40 acres, including an approximately 20-acre quarry pit and associated areas that will be built out as sediment is deposited and contoured.

It is estimated that the project will be implemented over the next 10–11 years and will be constructed in five phases with restoration of each phase implemented sequentially. The project assumes an annual sediment disposal budget of approximately 100,000 yd³/year for a total deposition of roughly 1,000,000 yd³. Sediment sources include the Goat Canyon sediment basins and channel maintenance, including the maintenance of Smuggler's Gulch and the channels of the Tijuana River.

METHODS AND RESULTS

Total impacts to 14.4 acres of Diegan coastal sage scrub (DCSS) will be mitigated at a 1.5:1 ratio through creation of high quality DCSS implemented during the phased project. DCSS species will be planted using one-gallon container stock and seed. Each phase will be irrigated as implemented until plants are established. A qualitative five-year monitoring program will be conducted with the success criteria targets of 80% total cover and 85% species diversity.

LESSONS LEARNED

There is some interest by land managers in the river valley that they be allowed to use the site for disposal of sediment excavated for restoration projects. If such a use were allowed, lessons learned from the Fenton Quarry restoration would be directly applicable, particularly the need to cap saline soils with native upland soils and the potential need to use gypsum to control salts that may wick to the surface.

22 FIVE-ACRE RESTORATION PROJECT IN BORDER FIELD STATE PARK

Lead Agency:
International Boundary and Water Commission (IBWC)
Project Administrators:
URS Corporation, SWIA, Ecological Conservation Management, ACS Environmental
Restoration Ecologist:
Boland Ecological Services
Dates: 2011–2016
Funding Agency:
IBWC
Reported in:
Progress Report: Five-acre Restoration Project in Border Field State Park. Year 2: September 15, 2012–September 14, 2013. John Boland and Mayda Winter. Southwest Wetlands Interpretive Association (SWIA).

This project was required of the IBWC as partial mitigation for impacts associated with upgrades to the International Wastewater Treatment Plant. The project involves restoring five acres of an old agricultural field within BFSP to a self-sustaining, perennial, native plant community that meets several success criteria. A restoration plan for the site, developed by John Boland in 2011, was accepted by the USFWS and called for a five-year restoration project that involved the control of invasive plants, revegetation with native plants, and monitoring. The restoration was to begin in 2011 and was to be fully implemented by December 2016.

The project site is located within BFSP and is five acres in size (Figure 38). The site was an old agricultural field that had been left fallow for approximately 30 years. It was formerly within the Goat Canyon watershed but in 2005, when the Goat Canyon sediment basins were constructed, the project site was cut off from most flows and so is relatively dry.

The dominant plants in the project site at the start of the project were a mix of agricultural weeds, including black mustard (*Brassica nigra*), giant reed (*Arundo donax*), and native riparian plants, such as mule-fat (*Baccharis salicifolia*). Giant reed had been treated several times during the preceding decade but a few stands were still alive at the start of the project.

Under subcontract to the URS Group, SWIA began the restoration project in fall 2011. SWIA hired two consulting firms to do the field work during Year 2: Ecological Conservation & Management, Inc. (ECM; prior to September 15, 2013) and ACS Environmental (ACS; post-September 15, 2013).

During Year 1, invasive plant species were treated and more than 2,200 native plants were installed and irrigated with drip irrigation. Native species included mule-fat, California encelia (*Encelia californica*), coastal sagebrush (*Artemisia californica*), coast goldenbush (*Isocoma menziesii*), and California buckwheat (*Eriogonum fasciculatum*), among other species.

The site received supplemental irrigation until November 2014. Thus, the 2015 growing season was seen as a critical test of the sustainability of the site.

Dr. John Boland conducted monitoring of the restoration project from 2011 through 2016. The monitoring followed the progress of the restoration and determined whether the project met the success criteria in the restoration plan. Qualitative and quantitative monitoring were conducted as described herein.

METHODS

Qualitative Monitoring. The overall appearance of the restored site through December 2015 was assessed using photographs taken from five permanently established locations.

FIGURE 38.
FIVE-ACRE RESTORATION PROJECT SITE (GREEN BOUNDARY)
SHOWING THE LOCATIONS OF THE WATER SOURCE (W), PHOTO POINTS (P)
AND VEGETATION SURVEY AREAS (S)

Quantitative Monitoring. Changes in the structure of the vegetation were documented by monitoring the vegetation along five 25-meter transects. Along each transect, the percent cover of each species was determined using the line-intercept method. The vegetation survey sites had been chosen in a stratified-random manner.

In order to determine the success of the planted container plants after the irrigation was turned off, the growth (height) and survivorship of 128 plants that were planted during winter 2011–2012 were monitored.

RESULTS

By the end of 2015, the project had achieved the following:

- Successful treatment of target invasive plant species;
- Successful vegetation with native plants;
- Attainment of project success criteria for the first and second years.

Results of the line intercept analysis conducted along the transects indicated that the restoration is successful. The targeted invasive species have been successfully treated and have not re-established and the native planted at the site have expanded since the start of the project. The plants had a high survival rate following the termination of supplemental irrigation November 2014. Of the 128 plants monitored, only four died (97% annual survivorship). The plants we planted are now well-established and likely to be a permanent feature of the vegetation for many years. Most of the plants are reproducing produced flowers and seedlings. The plants in the project site are therefore functioning as a natural community.

LESSONS LEARNED

Like the Tijuana River Estuary Water Quality Improvement and Community Outreach Project, this project demonstrated that areas of invasive, non-native species, introduced through sediment deposition and other disturbances, can be effectively controlled and revegetated with native species.

23 BUNKER HILL REVEGETATION: BORDER INFRASTRUCTURE SYSTEM

Lead Agency:
> U.S. Customs and Border Protection (CBP)

Project Administrators:
> U.S. Customs and Border Protection/HDR/RECON

Restoration Ecologists:
> HDR/RECON

Funding Agency:
> CBP

Dates: 2012–present

Reported in:
> 120-Day Installation Progress Report. U.S. Customs and Border Protection, Bunker Hill Revegetation, Border Infrastructure System, San Diego County, CA.

> Year 3 Report for the Smuggler's Gulch Revegetation: Border Infrastructure System, Area V., San Diego County, California. Dudek. October 14, 2014.

As presented previously, the Department of Homeland Security filled Smuggler's Gulch with over 2,000,000 yd³ of earth as part of a new triple border fence and high-speed roadway constructed between San Diego and Tijuana termed the Border Infrastructure System (BIS). In addition, a hillside to the west of Smuggler's Gulch, known as Bunker Hill, was highly disturbed during construction. Although initially reluctant to vegetate the bare earth exposed by grading and filling (because of exemption for mitigation granted by the waiver for BIS construction), the CBP agreed to initiate revegetation after pressure from resource agencies and environmental groups. In 2012, the CBP retained HDR to develop and implement a revegetation plan for the Bunker Hill area using native upland species. The Bunker Hill project area that was revegetated was approximately 5.6 acres in area and an additional 1.8 acres of road closures and a utility trench were also treated. RECON Native Plants (RNP) stored plants salvaged from the BIS at their nursery in the Tijuana River Valley for replanting and propagated additional container stock as part of the revegetation plan. The plan originally also included the Goat Canyon staging area of the BIS but was later deferred to fall of 2013. Two other small road closure areas were revegetated by RNP using plants supplied by BFSP. The initial planting was finalized in June 2013 and the results of the 120-day installation period was submitted September 2013.

The primary goal of the plan was to establish native perennial cover in areas disturbed by construction of the Border Infrastructure System and prevent or limit the establishment of non-native invasive plants.

METHODS

The revegetation plan addressed a number of planting zones (Figures 39 and 40). Soils were ripped to a depth of 12 inches in zones 1, 4b and 4d due to compaction in these areas. Soil roughening to a depth of one inch was conducted at zone 5. Cut and fill slopes comprising zones 2 and 3 were not suitable for soil ripping or roughening. Hydroseed slurries consisting of native seed and soil amendments were applied to zones 1, 2, 3, 4b, 4d and 5d in January 2013. Zones 5a, 5b, 5c and 5e were hand seeded, raked and covered with three inches of cedar bark.

Salvaged plants, including 11,607 plants of 32 different species, were stored for replanting at RECON nursery. Of these, 8,348 survived and were replanted at a density of 1,150 plants/acre over the 5.6-acre portion of the site representing zones 1, 2 and 3.

Supplemental irrigation was supplied in the form of overhead irrigation. Plants were watered once per week with approximately 40,000 gallons/acre.

Non-native invasive plant species were treated with glyphosphate and the biomass removed. Herbicide treatment was conducted April–May 2013. Erosion control measures, including fiber rolls and silt fences, were used in all areas. Some areas also were covered by jute netting and sand

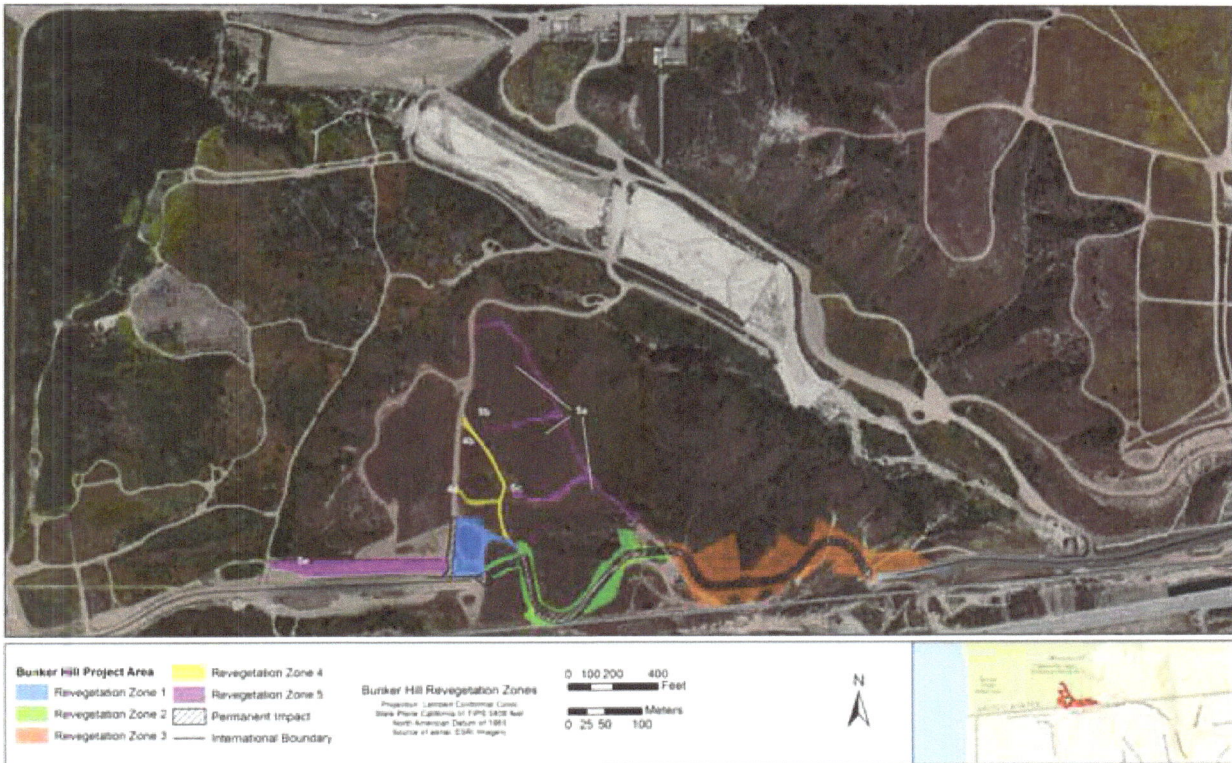

FIGURE 39. BUNKER HILL REVEGETATION ZONES

bags were employed where surface erosion was detected.

Monitoring was conducted monthly and included observations of plant health, inspection of irrigation systems, and effectiveness of erosion control measures. A plant mortality census was conducted in May 2013. Twelve random points were sampled during this survey: six in zone 1, four in zone 2 and two in zone 3. At each point, 20 plants closest to the point were counted, identified to species and noted as live or dead.

RESULTS

The results of the mortality survey indicated an average mortality of 4.29%. Thus, more than 95% of salvaged and propagated plants survived the 120-day establishment period. Erosion was noted in some areas but was considered to be minor. Some areas of hydroseed and hand seeded sites did not germinate and were essentially bare ground. Based on these results, a number of recommendations were made, including:

- Gradually reduce watering to accustom plants to no supplemental irrigation;
- Continue to treat and remove non-native invasive species;

FIGURE 40. REVEGETATION ZONES 4A AND 4C

- Continue to monitor erosion annually;
- Continue to monitor mortality monthly during the growing season (November–April);
- Conduct revegetation activities originally planned for the Goat Canyon staging area;
- Determine if rehydroseeding is required in areas that did not germinate;
- Reseed hand-seeded areas.

In May 2014, quantitative monitoring of canopy development was conducted using the point intercept method along 24 transects. Mean cover was 46.7% exceeding project goals for year 3 (2014) by 6.7%. Density of native species nearly doubled in 2014 relative to installation (2012) with colonization by some species and reproduction of others.

LESSONS LEARNED

This project, like the Tijuana River Valley Regional Park Trails and Habitat Enhancement Project, entailed cooperation by the CBP to close and revegetate portions of unofficial roads and areas disturbed by the BIS project. Over 11,000 native plants representing a number of species were salvaged and held at a nursery for the restoration. Salvage of existing plants takes advantage of an existing resource in restoration and can lower costs.

STUDIES OF RIPARIAN COMMUNITIES IN THE TIJUANA RIVER VALLEY

JUST AS DR. ZEDLER, University of Wisconsin, Madison and her associates conducted experimental manipulations of salt marsh associated with the Model Marsh and Tidal Linkage projects, so has Dr. John Boland of Boland Ecological Services expanded his exotic species removal expertise into active and passive restoration techniques of riparian habitats. These were recently summarized in the scientific literature, as presented herein.

Factors Determining the Establishment of Plant Zonation in a Southern California Riparian Woodland. John Boland. Boland Ecological Services. Madroño. Vol 61. No. 1, pp. 48–63, 2014.

METHODS AND RESULTS

This paper describes plant zonation in a Southern California riparian woodland and identifies the factors responsible for the zonation. The study:

- Describes the distribution of adult plants within the community;
- Describes the distribution of seedlings in areas of new recruitment;
- Examines factors affecting recruitment —timing of fruiting, timing of water level decline, and timing of seedling establishment—to determine their influence on adult zonation;
- Examines seedling survival and change in community structure over time to determine the influence of these post-

recruitment factors on adult zonation; and

- Discusses how these findings can improve riparian restoration projects.

Boland followed three discrete riparian forest segments within the Tijuana River Valley that developed after floods removed nearly all existing vegetation in 1980, 1993 and 2005. Thus, the forest segments were 32, 19 and seven years old, respectively, in 2012. Of the 25 tree and shrub species that are common to the Tijuana River Valley, three were numerically dominant: mule-fat (*Baccharis salicifolia*), arroyo willow (*Salix lasiolepis*) and black willow (*Salix gooddingii*). Adults of those species displayed significant elevational zonation, with mule-fat, arroyo willow and black willow most abundant in the high, intermediate and low zone, respectively. Among new recruits, arroyo willow and black willow seedlings displayed zonation similar to that of adults, indicating that zonation of these species was established at the time of recruitment. In contrast, mule-fat seedlings were more broadly distributed than adults; they were abundant in all zones, particularly in the low zone where adults were rare, indicating that zonation of mule-fat was established post-recruitment.

For black willow and arroyo willow, the timing of fruiting and the timing of water level decline were factors producing adult zonation. The two species had nearly non-overlapping periods of seed production, which led to zonation of their seedlings on the bank as water levels declined. The seedling zonation was then retained in the adults. For mule-fat, the seedlings of which were widely distributed, zonation of adults was the result of poor seedling survival in the low zone during the first winter and poor adult survival in the intermediate zone later.

LESSONS LEARNED

The results of this study can help guide future riparian restoration projects in Southern California. Based on the prolific natural recruitment of these dense, native-dominated stands, use of a natural restoration approach where possible is recommended instead of the more common horticultural approach.

Currently, Dr. Boland is assessing the effects of the shothole borer beetle on the riparian vegetation in the river valley.

SUMMARY OF LESSONS LEARNED

THIS CHAPTER summarizes the lessons learned from restoration efforts in the Tijuana River Valley and estuary according to biological and physical factors, and regulatory requirements. It is not intended to be a complete treatise of restoration techniques nor a "how to" guidance for tidal wetland restoration. Aspects of the topics presented herein are covered in much greater detail in *"Handbook for Restoring Tidal Wetlands"* (Zedler et al. 2001) and readers interested in a more comprehensive approach are referred to that work.

Lessons are divided into three different "factors." Biological factors include the propagation and establishment of wetland vascular plants. Physical factors include hydrology, soils and sediment scour and deposition. Regulatory factors include the conditions of discretionary permits and consultations required for restoration projects. All of these factors have been touched upon previously but each will be presented in greater detail here. Because habitat restoration is an ecological undertaking, biological factors such as plant propagation and growth are presented first, although it could be argued that these cannot be separated from the physical characteristics of soil and water. Regulatory issues are presented last and serve to illustrate the regulatory landscape that has evolved over the past 37 years.

BIOLOGICAL FACTORS

SALT MARSH VASCULAR PLANT ESTABLISHMENT

Early salt marsh restoration work at Tijuana Estuary focused on the propagation, establishment, and growth of low salt marsh dominated by cordgrass as this was identified as the preferred nesting habitat of the endangered light-footed clapper rail. Later projects included studies focused on restoration within the full range of salt marsh elevations including low, mid and high salt marsh, and the species that inhabit these elevational ranges. Methods for propagating and establishing these additional species were investigated and refined.

Prior to the restoration work conducted at Tijuana Estuary in the late 1970s and early 1980s, little was known about why cordgrass did not invade naturally into areas that appeared to have the appropriate elevation and substrate. As presented in the previous section, it was initially determined that cordgrass was a poor invader of sites with appropriate elevation and substrate as a result of its phenology, although that ultimately was refuted when cordgrass invaded the former sewage lagoons at Tijuana Estuary. Seed germination, as determined through laboratory experiments, was low, and naturally occurring seedlings had not been observed in the field. Field observations confirmed that cordgrass increased its distribution primarily vegetatively through growth of underground rhizomes that support above-ground stems. It was postulated that cordgrass increased its range through dispersal of clumps of individual plants that became dislodged and were transported to new sites by the tides where they became established. If true, the genetic diversity of cordgrass would presumably be quite low, represented by a single clone or a few clones. Unfortunately, little genetic work has been conducted on California cordgrass (*Spartina folisa*) other than to determine markers that differentiate it from other cordgrass species, such as smooth cordgrass (*S. alterniflora*), a native of the eastern U.S. introduced to northern California salt marshes by the USACE.

The hypothesis that cordgrass reproduced only asexually was disproven with the observation of cordgrass seedlings at Tijuana Estuary following winter flooding and reduced soil salinies in 1980. Such observations suggested that cordgrass did indeed reproduce sexually from seed, at least under certain environmental conditions.

Following field and laboratory experiments on plant establishment from seed, "sprigs" and "plugs," it was concluded that the most successful method of establishing cordgrass was through transplantation of plugs containing multiple aerial shoots with rhizomes with native soil attached. The advantages of transplanting rhizomes encased in native soil reduced transplant shock and allowed better initial rhizome growth as the native soils are often of higher quality than soils of the newly restored site. The disadvantages of transplanting plugs include cost and impacts to the donor site. Excavating relatively large plugs, transporting them to the restoration site and transplanting them is several times more costly than harvesting and planting a greater number of bare roots, or sprigs, per unit effort. Excavating the larger plugs disturbs a greater area of the donor site than harvesting the smaller bare root sprigs, although it has been documented that donor sites recover quickly.

Typically, the above-ground aerial shoots of transplanted plugs and bare root sprigs die back shortly after planting leading to the premature conclusion that the transplants are dead. However, the rhizomes usually survive and produce new aerial shoots the following growing season. After this initial growth phase, growth and expansion can be nearly exponential leading to rapid, often dense canopy development.

More recently, restoration projects at other Southern California estuaries and lagoons have examined establishment of cordgrass from methods other than plugs. Restoration of Batiquitos Lagoon in the late 1990s utilized bare root sprigs of cordgrass in the low marsh. Although initial mortality was relatively high, the surviving ramets expanded vegetatively and the typical distribution of cordgrass as the low marsh dominant was established within approximately five years after planting.

In 2010–2011, the South San Diego Bay Coastal Wetland Restoration and Enhancement Project restored approximately 230 acres of former salt evaporation ponds, known as the western ponds, to subtidal, intertidal mudflat and intertidal salt marsh habitats. The project originally called for planting the 52 acres of low marsh habitat with 56,874 nursery-grown cordgrass propagated from seed. This decision was based on the restoration team's desire to minimize impacts to existing cordgrass donor populations. However, more than seven months after seeds were collected and germinated, the contracted nursery was only able to deliver approximately 7,000 mature individuals. The low yield of plants propagated from seed required a change in strategy for cordgrass establishment. It also offered an opportunity to compare the survival and growth of cordgrass using three propagation and transplant methods and the costs associated with each.

Cost estimates for establishing cordgrass from plugs exceeded the planting budget and it was decided that the majority of the low marsh would be planted with bare root ramets. A total of 31,900 cordgrass individuals were planted as "bare root planting units" defined as a ramet of two to three aerial stems of cordgrass with two to six inches of rhizome with a minimal amount of native soil attached to the rhizomes.

In order to compare the effectiveness of establishing cordgrass grown from seed to those harvested and transplanted as bare root units or plugs, a randomized block experiment was conducted. Ten 60-foot x 60-foot (3,600 ft²) randomized blocks were established for each treatment and one equal

size unplanted control plot. Within each 3,600 ft² plot, 100 cordgrass individuals were planted on six-foot centers from nursery-grown stock, harvested plugs and harvested bare root sprigs. Thus, each of the 10 study plots included equal size randomized blocks of the three propagation methods with equal numbers of each treatment (n = 1,000).

One year after planting, monitoring of survival of cordgrass within each block was conducted. Survival of plugs was approximately 38% (SD 11.6%); survival of nursery-grown plants was approximately 33% (SD 19.7%); and survival of bare rooted plants was approximately 3% (SD 4.4%; Nordby Biological Consulting [NBC] and Tijuana River National Estuarine Research Reserve [TRNERR] 2013). As expected, cordgrass did not establish in control plots. Two years after planting, percent cover by cordgrass was visually estimated. Mean estimated cover of plugs was 12% (SD 6%); mean cover of nursery-grown plants was 13% (SD 10.7%); and that of bare roots was 2.9% (SD 4.4%; NBC and TRNERR 2014). Visual estimates of cover in 2014 yielded mean values of 25.8% (SD 13.1%) for plugs; 25.1% (SD 13.3%) for nursery-grown; and 5.3% (SD 7.43%) for bare root (NBC and TRNERR 2015).

Based on the results of the randomized block planting experiment, bare root sprigs are not a cost-effective method of establishing cordgrass. Initial survival was low and mean cover after three years was considerably less than other methods. It should be noted that bare root plants that were planted in areas other than the randomized plots appeared to have had better survival and growth. However, there are no data on these areas.

Conversely, nursery-grown cordgrass plants are a cost effective propagation method for salt marsh restoration projects; however, considerable lead time and, potentially, contracts with multiple nurseries would be required for large-scale efforts. Viable cordgrass seed comprises a small fraction of the flowering culms. An analysis of the percent live seed collected for the salt ponds project conducted by a certified seed laboratory concluded that live seed comprised 5.49% of the total collected with 11% germination. Thus, large quantities of seed must be collected and large areas devoted to germination trays. Once the cordgrass has germinated and rooted, it can be split in ramets with each split doubling the number of plants. During the seven months that the salt marsh plants were held at the nursery, only 7,225 cordgrass plants were propagated using the germinate and split method and 7,000 of these were deemed healthy for transplanting. Therefore, a lead time of one to two years would be required to produce the quantities necessary for large-scale restoration, a time frame that is not feasible in many cases. However, contracting with multiple nurseries could reduce that time frame. In terms of cost to the project, each plant delivered by the nursery cost $2.00 compared to a range of $5.75–$9.13 for collection and installation of each plug as contracted to the firm that installed them for this project.

Successful establishment of mid- and high salt marsh plants from plugs or cores containing roots and native soil has also been documented, most often from plants that were within the restoration footprint and were salvaged and replanted. Such salvage reduces the impacts associated with restoration and can be an economical method of procuring mature plants, especially if the contractor has mobilized large excavating equipment, such as front loaders and excavators. "Sods" of mid- and high salt marsh grasses, such as saltgrass (*Disitichlis spicata*) and shoregrass (*Monanthochloe littoralis*), can be excavated with front loaders and transplanted immediately or held in water-tight containers and watered for several months before installation. Excavation and immediate replanting were conducted for the Model Marsh project while the latter method was employed for the Tidal Linkage project. As demonstrated in the Tidal Linkage project, blocks of mid- marsh species can be excavated using a custom made bucket and small excavator.

Some projects do not impact existing salt marsh or do not include salvage due to added cost. Most of these projects include mid- and high marsh species grown from seed or cuttings. Some mid- and high marsh species, such as Bigelow's pickleweed (*Salicornia bigleovii*), arrow grass (*Triglochin maritima*), Parish's pickleweed (*Arthrocnemum subterminale*), alkali weed (*Cressa truxillensis*) and the endangered salt marsh bird's beak (*Chloropyron maritimum* ssp. *maritimum*) establish well from seed and can be grown readily in a nursery. Others, such as sea-blite (*Suaeda esteroa*), saltgrass, shoregrass, sea lavender (*Limonium californicum*) and alkali heath (*Frankenia salina*) are more commonly propagated from cuttings. Both cuttings and seeds should be collected from marshes in the same geographic region as the restoration site to preserve genetic integrity.

Most plants are grown in small (2.25-inch x three-inch) "rosepots" although some are grown in one-gallon containers. The smaller plants grown in rosepots typically experience less transplant shock and begin initial growth sooner than those in larger containers.

All plants grown in the nursery should be "hardened" prior to delivery and installation. Under salt hardening, all plants are subjected to a sea water tolerance program that entails gradual adaptation of each species to increasing salt content until all plants can tolerate irrigation with full strength sea water (approximately 34 parts per thousand NaCl) for a period of not less than 30 days. Under sun hardening, all plants grown in shade or semi-shade are subjected to full sun for a similar period prior to planting. Hardening dramatically reduces transplant shock and increases survival.

The ideal time for planting salt marsh restoration projects is winter. Lower temperatures result in less stressful conditions for transplanted nursery stock and salvaged plants, seasonal rainfall provides natural irrigation and high tides typically reach portions of the high marsh.

Planting in spring or summer results in temperature and water stress.

The cost to grow and install each rosepot container stock is approximately $2.00, and combined cost to grow and install is approximately $4.50, depending on the contractor. Therefore, the budget for restoration projects may affect the density of planting in project design. Planting density considerations are presented in more detail herein.

As Dr. Zedler and associates demonstrated in their experiments associated with the Tidal Linkage and Model Marsh projects, as well as long-term monitoring in the natural marsh at Tijuana Estuary, Pacific pickleweed (*Salicornia pacifica*) and salt marsh daisy (*Jaumea carnosa*) become co-dominant at the expense of shorter lived, less aggressive species, such as Bigelow's pickleweed and estuary sea-blite. Pacific pickleweed and salt marsh daisy readily establish from seed and form dense canopy cover that precludes germination and growth of other species. As a result, it is recommended that these species be excluded from the planting palettes of future restoration projects at Tijuana Estuary. Re-establishment of species diversity should be a high priority restoration goal.

The density of planted marsh species will affect the timing of canopy development which is often a success criterion of restoration projects, especially those that are intended as mitigation. A healthy natural salt marsh often exhibits canopy cover of 90% or greater and comparison with reference sites is a common requirement of agency permits. In general, the more densely a site is planted, the sooner it will develop a mature canopy. The trade-off is cost. As described previously, salvage and planting of cordgrass plugs can be costly, as can be plants grown in a nursery. The size of the restoration site can also dictate planting density. Smaller sites can often be planted at greater densities than larger sites.

Typical planting densities employed in restoration projects at Tijuana Estuary have included plants spaced at one-meter intervals on center (oc); two-meter intervals oc; and, on 2.5-foot, three-foot or six-foot centers. For example, for the Tidal Linkage project, cordgrass plugs were planted three-feet oc and salvaged mid-marsh species were planted 2.5-foot oc. The project achieved the canopy cover objectives after three years and monitoring was terminated. An example of marsh development over time is demonstrated by monitoring of Transect A, which exhibited 94% of point intercept hits as bare ground in Year 1; 59% bare ground in Year 2; and 14% in Year 3.

The Model Marsh included an experiment on planting densities for cordgrass. Cordgrass plugs were planted in treatment plots at two-meter and four-meter oc, with and without soil amendment in the form of processed kelp. Areas between treatment plots were planted at one-meter oc. As presented previously in Figure 37, cordgrass canopy cover after 20 months increased rapidly for most treatments. Amended two-meter oc plantings and unamended one-meter oc plantings achieved greater than 50% cover after this time period. Amended four-meter oc plantings achieved 50% cover, while unamended two-meter oc and unamended four-meter oc plantings achieved approximately 30% and 20% cover, respectively. Thus, the addition of processed kelp clearly accelerated cordgrass growth. The ability to amend soils with processed kelp in future restorations is restricted by the closure of the kelp harvesting operations in the Southern California region. However, looking at just the unamended data, it can be seen that canopy cover of unamended one-meter oc plants was approximately twice that of unamended two-meter oc plantings and three times that of unamended four-meter oc plantings. Over time, all plantings coalesced into a dense, nearly monotypic cordgrass marsh.

The Model Marsh was not constructed as mitigation for another project's impacts and, thus, did not have a canopy structure success requirement as a permit condition. Planting at four-meter oc would have likely sufficed but would have taken considerably longer to develop a dense canopy. With the coverage of the combined planting densities, light-footed Ridgway's rails inhabited the site in 2003, approximately three years after planting and monitoring was terminated.

The Napolitano Restoration Project at Tijuana Estuary provides an example of densely planted salt marsh species and soil amendment with the goal of achieving success criteria quickly. This project was constructed as partial mitigation for impacts to San Diego Bay from storm drain discharges. Caltrans District 11 was willing to plant densely to achieve permit conditions rapidly. Soils amendment was unanticipated and was conducted in response to the discovery of subsurface contamination by waster asphalt.

In all, 7,523 plants, including 3,000 cordgrass plugs, were planted three-feet oc. This represented an overplanting of approximately 37%. The restored marsh developed rapidly, with areas dominated by Pacific pickleweed demonstrating a mean of 87% cover by the end of Year 2 (n = 3 transects). Areas dominated by cordgrass had a mean cover of 69% (n = 2 transects) by the end of Year 2 and the single transect located in high marsh dominated by saltgrass demonstrated 71% cover. These coverage data exceeded the coverage data at the reference transects by the end of Year 2.

Lessons learned from the Tidal Linkage Project, as well as the Model Marsh and Napolitano projects include the need for supplemental irrigation for high marsh that is rarely inundated by the tide. As soils are exposed and dry, hypersalinity can add to the stress of the planted species. Supplemental irrigation increases root development during the establishment phase and decreases salinity stress. It should be noted that irrigating often increases invasion by weed species that can be difficult to control once established. Discretionary permits issued by regulatory agencies for mitigation projects typically include conditions that the site be self-sufficient, e.g., no maintenance activities

other than weed and trash removal, for three years prior to sign-off.

Projects that include irrigation should provide adequate funding for removal of all irrigation lines, sprinklers and controllers. Too often, these are left in place following the end of long-term monitoring programs. An example is the Model Marsh where the main irrigation line, controllers and sprinkler heads are evident today, 18 years after construction.

In applying the lessons learned from restoration projects at Tijuana Estuary to other regional wetlands, it is important to understand local tidal conditions and their effect on salt marsh plant elevations. Many southern California lagoons and estuaries, including Tijuana Estuary, develop shoaling or sills at the tidal inlet. High tides usually mimic those of the open ocean; however low tides may be moderately to severely truncated such that system does not completely drain even on a negative tide. This can affect the local distribution of low, mid- and high marsh. In another example, high tides within large embayments, such as San Diego Bay, can be higher than those on the open coast due to harmonic waves that resonate in the bay. Thus, elevations of low, mid- and high marsh are shifted upwards compared to other systems.

RIPARIAN VASCULAR PLANT ESTABLISHMENT

Establishment of riparian vegetation is dependent primarily on the depth of the groundwater at the restoration site. If the groundwater is near the surface, supplemental irrigation is not necessary and successful establishment is likely. If the groundwater is well below the ground surface, irrigation will likely be needed for multiple years. Also, if there are drought years, most riparian restoration projects will require irrigation.

One of the unique features of many Southern California riparian trees and shrubs is their ability to root from cuttings. Willow (*Salix* spp.) and mule-fat (*Baccharis salicifolia*) cuttings can be harvested from existing plants and planted either vertically or horizontally. With sufficient moisture, they root and grow aerial shoots. Thus, trees and shrubs for restoration are virtually free, requiring only the cost of labor to collect and install cuttings. Willow cuttings are generally four to six feet long and 0.25 to 0.5 inch in diameter with approximately one half of the length buried in the ground. Longer cuttings can be used if groundwater levels are known and a particular depth must be reached. To preserve the health of the donor plant, usually no more than 10% of the existing branches should be harvested.

Restoring a species-rich riparian habitat also requires shrubs and annuals. These typically are established from container stock or seed. Like the tree species, the establishment of these plants is dependent on ground water levels. Water at or near the surface is ideal, while deeper groundwater levels necessitate supplemental irrigation.

Shrubs require cultivation, usually at a nursery. One-gallon containers can usually be obtained for $3.00 or $4.00 each. Smaller size containers can be purchased for around $2.00. Seed for many species is commercially available. Price varies by species.

The 6.73-acre restoration site for the South Bay Water Reclamation Plant and the Dairy Mart Road Improvements Project is an example of a riparian restoration project where the groundwater level was high. Although irrigation was installed it was not needed and the willow cuttings and one-gallon container stock had a high survival rate and 90% after three years.

By comparison, some of the work done by John Boland represents an example of the challenge of restoring riparian areas where irrigation water is not available and the water table is far below ground surface. Boland tried a number techniques to improve survival of planted cuttings, including addition of mulch in the planting basins and planting in areas shaded by existing trees. These met with limited success.

The Hollister Street Bailey Bridge Replacement Project is an example where groundwater levels were relatively deep and supplemental irrigation was employed to root the willow cuttings. Cuttings were successfully installed as vertical poles and bundles beneath rip-rap bank armoring.

NATURAL RESTORATION OF RIPARIAN HABITATS

Boland advocates restoration of riparian habitats by natural recruitment. He argues that horticultural restoration, where nursery-grown container plants are planted in low densities and farming practices, such as irrigation and weed control, are used to sustain the plants can be costly and not always successful; often there are extensive deaths when the irrigation is discontinued, and often the installed assemblage does not resemble a natural riparian community. In natural restoration, a site is prepared (usually cleared and graded), and revegetation is allowed to proceed naturally with little or no human intervention. Boland notes that this is an appropriate method in sites that are inundated by floods during winter and have natural seed sources nearby. Lacking inundation or a seed source limits the practicality of natural restoration. However, many of the riparian restoration projects presented in the previous chapter met these criteria and yet were restored using horticultural techniques.

Boland proposes that natural restoration is superior to horticultural restoration in four important ways:

- It produces a community with a high density and cover of seedlings and adults. If adult riparian trees are nearby, recruitment on the order of 40–1,500 seedlings per m^2 was observed. Over time, Boland predicts, naturally recruited stands will remain denser and

have greater cover than horticulturally restored stands.

- Natural restoration produces a community with the appropriate spatial distribution of species, including the down-slope zonation of dominant species. In horticultural restoration projects, species are often mixed in space resulting in poor survivorship of individuals planted at less than optimal elevations, especially once irrigation has been discontinued.

- Natural restoration results in a community in which local plant species have appropriate sex ratios and genetic diversity. Some horticultural practices, such as the use of multiple cuttings from only a few source individuals or use of non-local stock, can result in unnatural sex ratios or unnatural genetic diversity.

- Natural restoration is likely to be less expensive than horticultural restoration. It requires no container plants or workers to plant them, no irrigation systems or workers to install and maintain them, and no weed control or post-recruitment maintenance.

ESTABLISHMENT OF UPLAND HABITATS

Upland habitats can be established from nursery-grown container stock and seed, generally applied as a hydroseed mixture. Supplemental irrigation is almost always necessary for one or two years after planting.

Some nurseries maintain stock of certain upland plant species for purchase, or such species can be custom grown from seed. One-gallon containers are around the same price as those for salt marsh and riparian restoration projects—$3.00 or $4.00 each. Seed can be collected from similar habitats in the vicinity of the restoration project. Many land managers require that seed be collected nearby in order to maintain genetic integrity of the species and habitats. Seed collected in this manner should be tested for purity and percent germination by a qualified seed laboratory prior to application, as some species may not produce viable seeds during a given growing season. The cost to collect and test seed varies, but an example of actual cost for an upland restoration in the Tijuana River Valley is presented here. Upland seed mixes that may have been collected from elsewhere in Southern California are available from seed companies. These seeds have been tested for purity of percent germination and are less expensive than collecting at the source and testing. For example, a well-established seed company in the region carries a diverse mix of Diegan coastal sage scrub seed mix that costs $45.00/pound. They recommend application of 51 pounds/acre for a cost per acre of $2,295.00. Seeds are typically applied as a hydroseed slurry, with seeds and an inert matrix sprayed on the restoration site. Commercial hydroseed companies reviewed on the internet advertised rule-of-thumb application costs of approximately $3,500.00/acre. Using this estimate, the cost to obtain and apply seed from a commercial source is approximately $5,795.00.

Both hydroseed and container stock usually require water for the first year or two. Potable water can be expensive and, many times, water is not immediately available and must be piped relatively long distances. Depending on the size of the restoration project, water can also be applied from a water truck, though this is not practical for large projects. In some cases where water is not available, the hydroseed does not receive supplemental watering and instead the project relies on rainfall for germination and growth. The success of this strategy, sometimes referred to as "spray and pray," is particularly uncertain during times of drought. Even with irrigation, hydroseeded sites must sometimes be re-seeded due to low germination and survival.

Restoration of Fenton Quarry entailed collection and testing of seed from 13 maritime succulent scrub plant species for the 4.3-acre area. Cost per acre in 2000 was approximately $2,800.00. Seed was applied as a hydroseed with stockpiled scrub mulch spread over selected locations to facilitate establishment of desired species and to re-introduce microorganisms. The site received supplemental irrigation for two years and was deemed successful after three years, although some remedial actions (gypsum addition) and replanting with container stock was required.

The Tijuana River Estuary Water Quality Improvement and Community Outreach Project, the Five-acre Restoration Project in Border Field State Park, and the Bunker Hill Revegetation: Border Infrastructure System Project all used container stock, sometimes in concert with seeding, to achieve restoration success. All were irrigated—the five-acre site with drip irrigation and the others with overhead sprinklers.

WILDLIFE

Few mitigation projects require post-construction monitoring of wildlife, although monitoring prior to and during construction might be required as part of permit conditions. Non-mitigation driven restoration projects often include monitoring before, during, and after construction (especially in the context of adaptive restoration). The use of restored sites by wildlife, especially threatened and endangered species, is one measure of success that is often overlooked by critics of restoration as mitigation. Those critics often cite evidence that vegetative cover and species diversity at restored sites falls short of reference sites. However, the almost immediate use of some of these restoration sites by listed and other species of special concern indicates the value of restoring wetland habitats despite this observed shortcoming. In addition, wildlife observations are sometimes unrelated to monitoring and instead are reported by reliable sources, such as the USFWS on Refuge lands. These observations have provided valuable data on the functions of restored sites and have, in some instances, modified or terminated regular monitoring to avoid impacting sensitive species. The following section presents the results of

wildlife monitoring of some of the projects presented in previous chapters.

Construction of the South Bay Water Reclamation Plant and associated realignment of Dairy Mart Road and construction of a new bridge over the Tijuana River resulted in impacts to 1.8 acres of disturbed mule-fat scrub and southern willow scrub that was occupied by up to six pairs of the state and federal listed endangered least Bell's vireo. Permitting agencies required mitigation in the form of creation of 6.73 acres of southern willow scrub and 3.18 acres of waters of the U.S. at the bridge site as well as 3.6 acres of willow habitat at an off-site location adjacent to Sunset Avenue west of Saturn Boulevard.

At the approximately 10-acre bridge site, least Bell's vireo recolonized the restored site within three years after construction and the permitting agencies accepted an early release of the project from its proposed five-year monitoring program. At the 3.6-acre off-site restoration, two vocalizing male least Bell's vireos were detected during the breeding season, although the full five-year monitoring program was performed.

Although small (~1.25 acre), the Napolitano site demonstrates the value of even small restoration projects for wildlife. Early in Year 3 of the proposed three-year monitoring program, the state and federal listed endangered light-footed clapper rail (now light-footed Ridgway's rail) were detected vocalizing in the restored site. As such vocalizations indicate advertisement for a mate, the monitoring program was terminated at the request of USFWS Refuges.

The Model Marsh was the subject of extensive monitoring, both from a regulatory and scientific perspective. Dr. Zedler and colleagues monitored the use of tidal creeks by fish relative to the marsh plain in replicate plots, as well as many aspects of vegetation development. TRNERR staff monitored wildlife use, among other metrics. As presented previously, the lower salt marsh dominated by cordgrass developed rapidly and by the fourth year of the proposed five-year monitoring program, up to four pairs of Ridgway's rails were estimated to occupy the 20-acre site. At the request of the USFWS, monitoring of canopy development in the lower marsh was terminated to avoid disturbing the rails.

The mitigation site for the Goat Canyon Enhancement Project totaled 25.76 acres, including mule-fat scrub (20.71 acres), southern willow scrub (2.59 acres), maritime succulent scrub (2.16 acres) and southern mixed chaparral (0.3 acre). Monitoring of the mitigation site consisted of quarterly monitoring of restored habitats and annual monitoring of the state- and federally-listed endangered least Bell's vireo. During Year 3 (2007) of monitoring, vireos began using the mitigation area for nesting. Prior to that year, all nesting activity occurred in areas adjacent to the mitigation area. Although the company conducting the monitoring requested early approval by the permitting agencies, monitoring was conducted for the full five-year period.

One of the goals of the Fenton Quarry Restoration Project was to restore habitat for the state-listed threatened coastal California gnatcatcher (*Polioptila californica californica*) and other animals associated with maritime succulent scrub. Originally proposed for five years, project monitoring was terminated after three years as a result of achievement of project goals, including use of the restored site by coastal California gnatcatcher. Gnatcatchers were observed foraging during several monitoring surveys during the five-year monitoring program. Once detected on-site, all vegetation surveys were conducted outside of the gnatcatcher breeding season. On-site breeding was not confirmed.

PHYSICAL FACTORS

HYDROLOGY

Creating the proper hydrology at a wetland restoration site is critical to project success, especially in restoration of tidal wetlands. Hydrology has been described as the driving force for wetland development and functioning (Mitsch and Gosselink 1986). Hydrology determines the elevation breaks in wetland habitats based on frequency and duration of inundation, which in turn affects salinity; soil development and transport, including scour and deposition; growth and dispersal of plants and aquatic animals; water chemistry, including dissolved oxygen and temperature; and many other processes.

Typically, tidal wetland restoration planning includes design of efficient channels and tidal creeks based, in part, on existing natural channels in the same system. These are then modeled using a variety of hydrodynamic models, and the models predict the inundation regime and, thus, the elevation breaks between subtidal, intertidal mudflat, and intertidal low, mid- and high salt marsh. This provides the basic framework for the restoration plan. Restoration of riparian habitats typically occurs within an existing creek or river system, although not always (see Tijuana River Valley Exotic Removal Program, previous section).

Hydrology both affects and is affected by sediment scour and deposition. While it is the water that transports the sediment, the deposited sediment affects future hydrologic functions. The impetus for much of the restoration and associated work at Tijuana Estuary is the deposition of sediment from upland sources in the channels and tidal creeks in the south arm resulting in loss of tidal prism and conversion of habitat types. Restoration under conditions where sediment is abundant and mobile imposes unique challenges as presented in the examples herein.

TIDAL LINKAGE

The Tidal Linkage Project was designed to connect two wetland areas in the northern portion of the estuary by excavating a channel through upland habitat between the two sites (upper intertidal portion of Oneonta Slough with tidal lagoons). The channel connecting the two wetland areas was not subjected to hydrodynamic modeling. It included several nearly sinusoidal curves with straighter

portions at each connection. The landward edge of each sinusoidal curve was designed to transition from intertidal mudflat to low salt marsh, which then transitioned to the mid- marsh plain. The low marsh area was planted with cordgrass and the mid- to upper marsh was planted according to whether it was a part of experiments by PERL or strictly restoration.

Initial monitoring indicated that the low marsh was developing as planned; however, higher than expected velocities within the channel began to erode the low marsh, essentially straightening the channel. The plants and soils of the low marsh area were eroded and the soils deposited in the channel, reducing its depth and affecting its use by aquatic organisms. Most of the low marsh area was lost from the system and a fragment remains today.

Model Marsh

The Model Marsh was designed as a kidney-shaped extension of Old River Slough in the south arm of the estuary; however, episodic deposition of sediment conveyed from upland sources by Goat Canyon Creek filled that portion of the channel and project designers were forced to move the project further east to another, smaller intertidal creek. The ability of this smaller creek to provide the tidal influence to accomplish the restoration was modeled and found to be adequate. An extension of the existing creek was added along the northern boundary of the Model Marsh. This provided the hydraulic

connection with the replicate tidal creek networks that were the basis of the PERL experiment comparing marsh development and use by fishes in areas with and without tidal creeks. The project was constructed and long-term monitoring was initiated.

In January 2002 a winter storm coupled with high surf resulted in substantial deposition of sediment in the main channel of the Model Marsh (extension of existing channel) and in the replicate tidal creek networks and on the marsh plain. Sediment-laden water flowing from Goat Canyon breached the protective berm around the south side of the Model marsh and deposited sediment at that location. PERL researchers requested that the main channel be dredged in order to provide adequate tidal flows to constructed tidal creek networks so that they could continue their research. A work plan was developed to gather information required in order to acquire the necessary permits to conduct the dredging. Tasks included:

- Characterization of sediment deposited in the marsh to determine its source, including eroding banks of the Model Marsh, sedimentation from upstream sources (Goat Canyon) or sediment carried from the river mouth during storm events.
- Hydraulic analysis to determine whether dredging the main channel would result in a significant reduction of sediment within the tidal creek networks and on the marsh plain; expected duration of enhancement of tidal flows by dredging the main channel; frequency of future

dredging to maintain such flows; and proposed depth of dredging and disposal of dredged materials.

- Proposed project approach, including equipment needed for dredging and disposal, staging and access routes, dewatering of dredged materials.
- Identification of CEQA requirements—need for a supplemental EIR or categorical exemption.
- Permitting — identify information required for review should permits from regulatory agencies be required.
- Cost estimate.
- Timing/schedule.

Funding was secured for the first task—characterization of the sediment deposited in the restored marsh. A number of transects were established across the flow path of Goat Canyon upstream of the Model Marsh, within the main channel of the Model Marsh and within the tidal creek that feeds the Model Marsh. While the study did quantify the amount of sediment deposited in these areas, it failed to identify the source, concluding that it was likely from three potential sources—upstream (Goat Canyon), erosion of banks of the Model Marsh and sediment from the river mouth during storms.

Given the ambiguity of the sediment characterization tasks, the additional tasks identified in the work plan were not undertaken and there was no additional modification of the hydrology of the Model Marsh. This particular project points out the need to anticipate such impacts and

have plans for remediation, including conditions in regulatory permits that allow for remediation. This is discussed in further detail in section 3.3 Regulatory Permitting.

Unlike many Southern California estuaries and lagoons, Tijuana Estuary has a history of a continuously open inlet with three exceptions. Inlet closure in 1984 resulted in catastrophic impacts to the flora and fauna of the estuary (see Introduction) some of which still have not recovered. A partial inlet closure in 2009 occurred when the inlet migrated some 600 meters south of its typical mid-estuary position and closed the tidal channels that feed the south arm, including the tidal channel that feeds the Model Marsh. Some tidal action continued upstream of this blockage, but it was feared that the Model Marsh would experience some of the same impacts that occurred in 1984. The USFWS provided funding to remove the sand deposited in the inlet in 2010. It is unclear whether a fully tidal south arm would have prevented the southward migration of the inlet or the recent closures at low tide; however, it is possible that a restored tidal prism in the south arm would have scoured the sand that eventually closed the south inlet.

Most recently (2016–2018), the inlet has been subjected to a series of full closures, likely due to coarse sediment from a beach nourishment project in Imperial Beach being pushed into the mouth by El Niño-induced sea level rise and large waves. The USFWS has been re-opening after closure to avoid negative impacts, although a large storm

during one closure caused flooding and remarkably rapid development of anoxic conditions (probably through stratification). This resulted in a mortality of shellfish and fish, including leopard sharks, and has made mouth management a top priority.

Implementation of large-scale restoration, such as that presented in the Tijuana Estuary–Friendship Marsh Restoration Feasibility and Design Study, completed in 2008, may help to prevent future mouth closures. As highlighted earlier, a primary goal of TETRP since its inception almost 30 years ago has been to restore tidal prism and recover the ability of the estuary to naturally maintain its characteristically open conditions.

ROLE OF FRESHWATER

Based on the response of cordgrass to winter flooding in 1980 (see Coastal Wetlands Management: Restoration and Enhancement), some managers and restoration practitioners have advocated that tidal wetland restoration projects have a direct hydrologic connection to a freshwater creek or river. As noted previously, the effects of freshwater inflows on tidal wetlands is not a simple, straightforward relationship. While flooding of the north arm of Tijuana Estuary in 1980 resulted in dramatic increases in height and biomass of cordgrass, the potential negative effects of providing a direct hydrologic connection to a newly created or restored site should be carefully considered. Newly restored sites are susceptible to erosion and deposition which would be exacerbated by direct stream flows into the marsh and channels. Even a few inches of deposition can change the functioning of the tidal creeks and channels and convert target habitat goals, i.e., low salt marsh to mid- salt marsh or high salt marsh to upland habitat. Prolonged freshwater flows can convert salt marsh habitat to freshwater marsh habitats. This was demonstrated at the San Diego River Estuary where short-term conversion of salt marsh to freshwater marsh resulted from prolonged reservoir drawdown and at Los Peñasquitos Lagoon where longer-term type conversion occurred. Los Peñasquitos Lagoon continues to be impacted by fresh water with conversion of salt marsh to brackish and freshwater marsh in the southeastern end of the lagoon. Habitat conversion has also occurred in the south arm of Tijuana Estuary where what were once nuisance flows from Yogurt Canyon have now become perennial flows and have converted a large area of salt marsh to freshwater marsh. Freshwater flows also provide a mechanism for dispersion of weed propagules that would not survive in a saline environment.

The benefits of episodic reductions in soil salinity can still be obtained without directly connecting a freshwater source to the upstream end of a tidal wetland restoration. At Tijuana Estuary, the Tijuana River flows to the ocean almost one mile south of Oneonta Slough where measurements of the responses to freshwater on cordgrass were measured. Intensive monitoring of water salinity in the river and Oneonta Slough by PERL researchers in 1995 showed that salinities in Oneonta Slough were heavily influenced by river flows during and after rain events. Similar results have been documented through intensive long-term monitoring of the western salt ponds restoration in south San Diego Bay and freshwater flows in the Otay River during and after rain events. As most restoration sites are located within the drowned valleys of rivers and streams, diverting all or a portion of those streams into a restoration site is not necessary and can result in undesired effects.

SOIL AND SEDIMENT

Typically, the most expensive part of any wetland restoration project is the earthwork, including grading and disposal of excavated soils. All of the tidal restoration projects at Tijuana Estuary have involved the lowering of upland areas to the elevation of the tides. Most have involved disposal of fill at upland sites using conventional equipment (dump trucks). One project, the Tidal Linkage, disposed of excavated soils via dredge slurry and pipeline, as presented previously.

There are two main alternatives for dealing with excavated soils: beneficial reuse (e.g., in the ocean, on the beach, or in an upland site such as a quarry), and disposal at an upland landfill. A third option, disposal at an EPA-approved ocean disposal site, is sometimes investigated, but to date has not been used for restoration projects at Tijuana Estuary.

With predicted sea level rise, land managers are turning to retaining soils excavated for wetland restoration on-site for resiliency to rising tide levels. Soils may be used to create transition zone habitats that can convert to intertidal habitats in the future. This disposal option requires that there be area available that is disturbed or of low ecological value so that there is no loss of wetlands or other valuable habitats that would require mitigation.

Hauling soils to an upland disposal site is usually the most expensive option, especially if the final disposal site is a landfill. Not only is there the expense of the trucks, fuel, drivers etc., but most landfills charge a "tipping fee" that was estimated at $5.00/yd^3 in the Tijuana Estuary–Friendship Marsh Restoration Feasibility and Design Study (Feasibility Study). Estimated costs for disposal of soils excavated for the 250-acre Feasibility Study in 2008 at the nearest upland site (Nelson Sloan Quarry) were 2.50/per yd^3. Costs for disposal at the farthest landfill site in Lakeside, California were $28.00/yd^3 plus a $5.00/yd^3 tipping fee. Those costs are dramatically reduced if the soils can be trucked or slurried directly to the surf zone.

In order to qualify for beneficial reuse, permits are required from the USACE, EPA and/or the RWQCB which enforces the EPA standards at the state level. The USACE and EPA jointly oversee the Clean Water Act (CWA) and require a Section 404 permit for any discharges of fill into waters of the U.S. The RWQCB requires a Section 401 Water Quality Certification as their action related to the CWA. Other agencies, such as the California Coastal Commission (CCC), regulate activities within the coastal zone.

The USACE requires that beneficial reuse be examined for coastal projects, including restoration. Detailed discussion of permits required for wetland restoration plans and the lessons learned from them is presented in section 3.3 Regulatory Permitting. The focus of this discussion is the disposal of sediments excavated during construction.

The USACE and EPA have established procedures for evaluating the suitability of soil and sediments for various disposal alternatives. The guidance document, titled *Evaluation of Dredged Material Proposed for Discharge in the Waters of the U.S.—Testing Manual* (referred to as the "Inland Testing Manual"), provides a set of tiered approaches to evaluate the most appropriate inland, beach, or nearshore disposal options (EPA/ USACE 1998). If ocean disposal is proposed, sediment testing would be conducted according to the Green Book (EPA/USACE 1991). These sediment evaluation studies are required in order to obtain a disposal permit from the USACE. Permitting is done in conjunction with the EPA and other federal and state agencies. In addition to federal testing requirements, the San Diego RWQCB has established waiver criteria in the San Diego Basin Plan for industrial or commercial reuse of dredged material (RWQCB 1998).

In order to re-use excavated soils to nourish beaches or the nearshore environment, they should be clean and similar in grain size to the receiving site. Soils must be tested for a suite of potential contaminants and must not exceed screening levels established by the regulatory agencies. The USACE's rule of thumb for beach nourishment suitability is that the material should have 80% of the sediment greater than 0.075 millimeters in size, and be compatible with the grain size, texture, and color of the sand at the proposed receiver beach. This rule was founded on the observation that most contaminants are bound to fine sediments, such as clays and silts. However, it has been applied to soils that are not contaminated as well. This "80/20 rule" was the basis for the Tijuana Estuary Sediment Fate and Transport Project presented previously.

COSTS FOR LANDFILL DISPOSAL

Soils of proposed restoration sites should be tested at a commercial soils laboratory during the planning stages to determine the potential for chemical contamination, grain size and nutrient levels. This type of an analysis is outlined in a Sampling and Analysis Plan (SAP) that is reviewed by the agencies that regulate sediment and soil disposal. A SAP includes the number and location of soil cores to be collected; compositing of cores or strata, if proposed; the analytes to be tested for; the approved testing protocols and the appropriate screening levels used to determine the extent of contamination, if any. The SAP may be reviewed by the appropriate regulatory agencies (USACE, EPA, RWQCB) prior to collection and analysis, or the project may proceed at risk with the understanding that additional collection and testing may be required prior to issuance of project permits.

Soil coring devices are typically employed to characterize the soils of the site. These are hollow coring devices, usually two to four inches in diameter that are driven into the ground by a truck-mounted rig, or in some cases, by hand auger. Cores are often collected to depths exceeding the proposed elevation of the marsh plain in order to determine if there are layers of coarse sand that can be beneficially re-used, such as in beach nourishment, while the finer grained soils more typical of marsh soils are backfilled. For example, a restoration site that might require the excavation of eight feet of soil to achieve the proper tidal elevation for the restored marsh might include soils coring to 20 feet below ground surface. In order to be cost effective, soils collected in cores at discrete depths are often composited, e.g. mixed together, and tested for contaminants. For the 20-foot deep corings described previously, discrete two-foot units could be collected in the field (10 two-foot cores per site) and composted for later testing in the laboratory, with a portion archived in case additional tests are needed. Using this method, if all of the strata are determined to be free of contaminants or are below screening levels used by regulatory agencies further testing would not likely be required. If, for example, the zero- to two-foot deep surface layer tested high for pesticides, as is frequently the case, smaller depth increments could be retested in order to more accurately depict the vertical extent of this contaminant. A typical suite of potential contaminants that might be tested for include:

- Total petroleum hydrocarbons (TPH)
- Metals
- Polynuclear aromatic hydrocarbons (PAH)
- Chlorinated pesticides
- Organochlorine pesticides (OCP)
- Total polychlorinated byphenyls (PCB)
- Phenols
- Phthalates

In addition, valuable information on soils of the proposed restoration site can also be tested in order to inform the restoration process. Such factors as total organic matter, pH, and nutrients, such as nitrogen and phosphorous, can help to identify potential deficits in the newly exposed soils.

The soils of newly restored wetlands are often different from natural salt marsh soils in that they lack the highly organic fine silts and clays that develop with tidal action. For example, the soils of the Model Marsh project had a loamy, sandy composition although the 1852 map of the estuary indicated that the project is located in an area that was formerly intertidal salt marsh. Still, the site rapidly developed dense low and mid- marsh cover, although experimental cordgrass plots that were amended with processed kelp developed dense canopy cover more rapidly than did unamended plots. Thus, amendment of soils with kelp can accelerate salt marsh establishment and growth. However, amending the soils of large restoration projects with processed kelp is impractical from a cost and availability perspective. Processed kelp purchased for the Napolitano restoration cost $22.00/

yd^3 in 1998 and the source, KELCO, a kelp processing company, no longer operates in the San Diego region.

Based on grain size, it was determined that the soils of the Model Marsh were not compatible with the beach or nearshore sands adjacent to the estuary. Soils were hauled to a disturbed upland site in Goat Canyon and were stockpiled for approximately one year. They were then used to fill and restore Fenton Quarry. Despite a four-foot thick cap using native upland soils, salts from the excavated marsh soils leached to the surface and impacted the seeded maritime succulent scrub plant species. Gypsum was applied to the surface of the worst areas to successfully neutralize high salinities.

Soils of the Tidal Linkage Project were deemed suitable for beach nourishment through the process outlined previously. Soils were excavated using conventional, land-based equipment, such as excavators and front loaders, and "fed" to a cutterhead dredge that was floated in shallow pit. The soils were then mixed with water to form a slurry which was pumped to the beach. Permit conditions require that the slurry be tested for fecal coliforms on a regular schedule. Coliform numbers exceeded the test maximum on at least one date although actual data could not be located for this report. It was postulated that ground water could have been contaminated from leaking sewer pipes that once carried effluent to sewage lagoons (now referred to as "tidal lagoons") located at the eastern terminus of the Tidal Linkage Project. Despite high counts, the project was completed using the beach as a re-use site.

Soils of the area proposed for restoration in the Tijuana Estuary–Friendship Marsh Restoration Feasibility and Design Study (250 acres in the south arm) were characterized as poorly sorted sand, fine sand, and silts typical of a riverine system. The heterogeneous nature of the soils did not meet the 80/20 rule. However, some pockets of sand suitable for beneficial use were identified. The project proposed a somewhat complex disposal plan that included disposal of appropriate sandy material on the dunes and beaches, use of finer materials to construct a protective berm, and multiple disposal options for fines at various landfills, abandoned quarries and at the EPA-approved LA 5 ocean disposal site. The presence of DDT and its metabolites from former agricultural practices further complicates the disposal scenarios. Additional sediment characterization will be required prior to construction of the first phase of the project.

Salvaging existing soils from wetlands that will be modified by restoration projects has been shown to impart several important biological components that can accelerate marsh development. These include bacteria, detritivores, fungi and mycorrhizae and, potentially, a seed bank that may help to establish a diverse flora or augment plantings. Stockpiling and reusing existing soils reduces cost as such soils do not have to be hauled off-site. Salvaged soils may be stored on-site for weeks or months, as was the case in the Tidal Linkage restoration. That project stockpiled marsh soils on-site for the approximately 3.5-month construction period. The soils were covered with a tarp to preserve moisture and the tarp was removed and the soils irrigated with freshwater on an as-needed basis. Even the small volume of soil salvaged for that project (~80 yd³), spread thinly on the marsh surface, was sufficient to increase the rate of canopy development. PERL researchers recommend that the top five to ten centimeters of soil be salvaged and stored separately from deeper soils and that the storage time be minimized to prevent loss of biotic components from both anoxic conditions and prolonged exposure to oxygen. Larger projects may be constructed in phases so that soils can be salvaged and reintroduced sequentially.

The Fate and Transport Project was undertaken to test the efficacy of the 80/20 rule. Sediment from the Goat Canyon sediment detention basins, consisting of roughly 50% sands and 50% fines, were trucked to the surf zone and dumped to allow waves to rework them. The project showed no effects on biological target taxa. However, although the sediment was tested for bacteria prior to deposition, monitoring indicated that concentrations of fecal bacteria, in particular *Enterococcus*, briefly exceeded single-sample public health standards. The concentrations of *Enterococcus* were highly correlated to the deposition of fine sediment. This event illustrates the need for careful testing of soils before they are used for beneficial purposes. In order to receive permission from the regulatory agencies to conduct a larger-scale program, such as phases of the 250-acre restoration identified in the Feasibility Study, a net benefit to the beach and nearshore environment must be demonstrated. Future pilot programs, if any, should focus on benefits rather than impacts.

Fertilizer

The use of fertilizers in salt marsh restoration projects should be customized to the soils of the restoration site. Soils of restoration sites located on dredge spoils or in highly mineral soils may have very low nutrient levels (Zedler et al. 2001). Adding fertilizer to such soils can accelerate development of above-ground and below-ground biomass to increase survival of transplanted species. Broadcasting dry fertilizer and rototilling it prior to planting is one method to increase soil nutrients; however, most of this would not benefit widely-spaced plantings. Unexpected effects of nutrient enrichment have included shifts in species assemblages and changes in herbivory. Plant species differ in their nutrient demands and efficiency, and fertilizing mixed-species stands can lead to changes in species composition, including invasion by undesired weed species. Fertilization may also increase herbivory by increasing nutrient levels in tissues.

Methods other than broadcasting and rototilling dry fertilizer may improve survival and growth of transplants, depending upon method of transplantation. Individual plants grown from seed or cuttings can be fertilized with slow-release fertilizers in the

transplant pit prior to planting the plant. Plants salvaged with native soils, such as cordgrass plugs, do not typically require fertilization. Soluble fertilizers can also be used, but are not available for as long a period. Recommendations for fertilizer types and quantities can be found in *Handbook for Restoring Tidal Wetlands* (Zedler et al. 2001).

Regulatory Permitting

Due to their ecological value and history of loss from human activities, remaining wetlands are heavily regulated by a variety of agencies, including federal, state and local entities, such as cities and counties. The role of the USACE, EPA and RWQCB in regulating discharges of fill and water to wetlands through CWA has been presented in the previous section. The California Department of Fish and Wildlife (CDFW) regulates non-tidal wetlands through its Section 1600 Streambed Alteration Agreement permit and the California Coastal Commission (CCC) regulates wetlands within the coastal zone under the Coastal Zone Management Act. The City and County of San Diego have regulations protecting wetlands within their boundaries.

Wetlands also support a large number of sensitive species, including state- and federally-listed threatened and endangered species. This, again, is primarily the result of habitat loss. Many restoration projects have potential effects on such species and require consultation with the USFWS and CDFW.

Wetlands often contain historic and prehistoric cultural resources that affect the design of restoration planning, as well as its implementation. At Tijuana Estuary, CSP regulates the cultural resources of Border Field State Park and the USFWS regulates cultural resources through their own cultural resources unit. Similarly, the U.S. Navy, which owns a portion of the estuary, has its own cultural resources division.

All of the projects presented in this document have required some or all of these permits/consultations. Acquiring so called "discretionary" permits is a major part of restoration planning and there are many lessons that have been learned in the regulatory arena from projects at Tijuana Estuary. This section presents some of the challenges of regulatory permitting and offers an after-the-fact assessment of the way in which those permits affected projects and how restoration planners can improve wetland management through innovative permit conditions.

Endangered Species Act(s)

The Federal Endangered Species Act (FESA) provides a program for the conservation of threatened and endangered plants and animals and the habitats in which they are found. The lead federal agencies for implementing ESA are the USFWS and the NOAA Fisheries Service, formerly NOAA National Marine Fisheries Service (NMFS). The USFWS maintains a worldwide list of endangered species, although enforcement of the act is limited to the U.S.

The law requires federal agencies, in consultation with the USFWS and/or the NOAA Fisheries Service, to ensure that actions they authorize, fund, or carry out are not likely to jeopardize the continued existence of any listed species or result in the destruction or adverse modification of designated critical habitat of such species. The law also prohibits any action that causes a "taking" of any listed species of endangered fish or wildlife. Likewise, import, export, interstate, and foreign commerce of listed species are all generally prohibited.

California was the first state in the country to protect rare plants and animals through passage of the California Endangered Species Act (CESA) in 1970. The FESA was passed by Congress in 1973.

The CESA states that, "all native species of fishes, amphibians, reptiles, birds, mammals, invertebrates, and plants, and their habitats, threatened with extinction and those experiencing a significant decline which, if not halted, would lead to a threatened or endangered designation, will be protected or preserved."

In California, the CDFW oversees CESA in collaboration with other organizations and agencies. As part of this process, CDFW serves in an advisory role to help offset potentially adverse impacts to rare, endangered, and threatened species.

According to the California Resources Agency, the California Endangered Species Act "generally parallels the main provisions of the Federal Endangered Species Act and is administered by the California Department of Fish and Wildlife." In addition, the CDFW designates certain species as Fully Protected Species. Under this designation, there is no take authorization and projects that may impact listed species are reviewed on a case-by-case basis.

Under the FESA, take of a listed species and the conditions under which a wetland restoration project may or may not proceed is determined through a formal consultation process culminating in a Biological Opinion (BO). In some instances, informal consultation can suffice. The CDFW regulates state-listed species through its own process, though its definition of take differs from the federal definition.

The most common limitation placed on restoration projects at Tijuana Estuary regarding threatened and endangered animal species is restriction on work during the breeding season. As the vast majority of threatened and endangered species at Tijuana Estuary are birds, any work that might disrupt their breeding is prohibited. Generally, most birds, including threatened and endangered species, breed and raise their young in spring, roughly between March and September. Some threatened and endangered species, such as the light-footed Ridgway's rail, have been observed to breed from early February through September. The resource agencies typically designate an official breeding season for each threatened and endangered species, although the actual breeding season can

be highly variable depending on a number of factors, including prolonged drought or rain, temperature (both extreme heat and cold) and abundance of prey. For the light-footed Ridgway's rail, the USFWS defines the breeding season as February 15–September 15; however, in recent consultations with the CDFW, that agency defines the breeding season as February 1–September 30. Because restoration at Tijuana Estuary occurs on a USFWS wildlife refuge or on CSP lands, the federal designation has been applied to consultations.

All of the tidal wetland restorations at Tijuana Estuary were constrained by the Ridgway's rail breeding season. All work had to be conducted between September 2 and February 14. The Tidal Linkage Project and Napolitano Restoration Project, both located in Oneonta Slough, were completed well within that time frame. The Model Marsh Project, however, nearly exceeded the allowable construction time, finishing on February 14, 2000 after the discovery and recovery of a culturally-significant shell midden near its western terminus. The midden was discovered by an archaeological monitor during construction in December 1999. A recovery plan was formulated and a small army of archaeologists was dispatched to recover the shells and associated remains and analyze the findings. This find and subsequent activity increased the cost of project and nearly caused the project to run into the beginning of rail breeding season.

In addition to constraints during the breeding season long-term monitoring of

two of the three tidal wetland restoration projects implemented at Tijuana estuary were curtailed due to the presence of Ridgway's rails. These included the Model Marsh and Napolitano projects. Such curtailment may be viewed as good and bad—good because the project was successful in that it attracted this endangered species; bad because it limits what can be learned about each restoration site. Although monitoring could conceivably be conducted outside of the breeding season, in both cases the USFWS requested that monitoring be terminated.

Another federal- and state-listed endangered species, least Bell's vireo, has been the subject of agency consultation on several riparian restorations in the Tijuana River Valley. Unlike the year-round resident Ridgway's rail, least Bell's vireo is a migratory species that breeds in riparian habitats in southern California. Historically, it was likely rare in the valley due to lack of riparian forest, but development of this habitat type with freshwater runoff from Mexico has made the Tijuana River Valley much more suitable (Safran et al. 2017). The bird usually arrives from southern Baja California, Mexico in March and April and departs by late July or early August. Occupied breeding habitat for species was impacted by the South Bay Water Reclamation Plant and the Dairy Mart Road Improvements Project. That project successfully restored/created habitat that supported this species.

The Goat Canyon Enhancement Project also resulted in impacts to occupied vireo habitat. In this case, consultation with the USFWS significantly delayed project approval and implementation. During this delay, a storm event in winter of 2005 washed sediment through the partially constructed sediment basins onto existing high salt marsh/salt panne habitat effectively converting it first to willow/mule-fat scrub then to weedy upland habitat.

The Goat Canyon Enhancement Project was initially designed to have three in-stream sediment basins. The sediment basins were designed at various locations within the flow path of Goat Canyon Creek. The placement of the basins in these locations resulted in impacts to multiple least Bell's vireo nesting territories. The project was subsequently redesigned to include off-line sediment basins in a disturbed portion of Goat Canyon; however, one vireo nesting territory was still impacted by the design. A draft EIR/EIS was made available for public review on October 12, 2001. A final EIR was issued December 21, 2001. A draft Biological Assessment (BA) prepared pursuant to Section 7 of the FESA and the Section 2081 of the CESA was also submitted to the USFWS and CDFW on October 12, 2001. A revised BA was submitted in August 2002. A request for formal consultation accompanied by the BA was made by NOAA to the USFWS on August 22, 2002.

On January 16, 2003, the USFWS submitted the BO. In early March 2003, the project went before the CCC for issuance of a Coastal Development Permit. The CCC approved the project with modifications to improvements to Monument Road.

Documentation of existing conditions required field surveys and mapping of vegetation communities and focused or protocol surveys for each threatened or endangered species with the potential to occur in the project area. Focused surveys for a number of species were required for the Goat Canyon Enhancement Project. These included focused surveys for least Bell's vireo, southwestern willow flycatcher and arroyo toad in 2000; for least Bell's vireo and arroyo toad in 2001; and focused surveys for coastal California gnatcatcher in 2001. A wetland delineation was conducted pursuant to Section 404 of the CWA. In addition to protocol presence/absence surveys for vireo, spot-mapping was employed to determine the breeding territories of pairs that were detected in the project area. Impacts to sensitive habitats, such as southern willow scrub, mule-fat scrub and mule-fat/elderberry scrub were mitigated at ratios greater than 1:1. For those three habitats that were occupied by nesting least Bell's vireo, mitigation was required at 5:1; for those that served as foraging habitat but did not support nesting vireos, mitigation was required at 3:1. A total of 0.8 acre of occupied vireo habitat was impacted by the project requiring 4.0 acres of mitigation in the form of mule-fat scrub. Mitigation for the project totaled 25.76 acres, including 20.71 mule-fat scrub and 2.59 acres of southern willow scrub. Long-term monitoring was described previously.

Construction of sediment detention basins began in the fall of 2004. The project was approximately 50% completed when

construction was halted in March 2004 for the breeding season of the least Bell's vireo and other sensitive bird species, as mandated by the BO. Construction resumed in September 2004. Unfortunately, heavy rainfall in October 2004 filled the basins with sediment prior to their completion. Much of the sediment from the October rains was subsequently removed and construction resumed in November of 2004. However, additional rains in November and December filled the basins with sediment once again. This sediment had not been removed when record level rains occurred in mid-January of 2005. Because the basins were already full, sediment passed through the basins and was deposited on the disturbed salt marsh/salt panne habitat located to the west. This single event resulted in the burial of approximately 15 acres of salt marsh/salt panne under as much as two feet of sediment. Additional area was buried during another storm on January 28–29, 2005, which produced approximately one inch of rainfall.

Although it may be easy, in hindsight, to partially attribute this loss of habitat on the protracted permitting process and breeding season restrictions of threatened and endangered species, it nonetheless serves as an illustration of good intentions resulting in undesirable and unexpected outcomes. The mandate of the USFWS to protect least Bell's vireo inadvertently led to loss of valuable habitat that supported another sensitive species, the state-listed endangered Belding's savannah sparrow (*Passerculus sandwichensis beldingi*). Belding's savannah sparrows populated the high salt marsh in the general area covered by sediment.

Breeding season restrictions, by their very nature, force restoration projects, as well as other projects, to be conducted during the winter rainy season. This can lead to project delays and increased costs, as illustrated previously. Other projects at Tijuana Estuary have suffered similar fates. The Fate and Transport Project is another example. For this project, sorted sediment was then transported from the Goat Canyon detention basins to the beach approximately 0.5 mile south of the Tijuana River mouth via a haul truck along a dirt road that serves as a horse trail to the beach. A total volume of roughly 45,000 yd^3 was deposited in the nearshore in two phases: 10,000 yd^3 during Phase I in November 2008 and 35,000 yd^3 in Phase II in October 2009. Work was restricted to the non-breeding season of the California least tern (*Sterna antillarum browni*).

Shortly after work began in 2008, a relatively small rain event made conditions on the dirt horse trail soft. Dump trucks sunk up to their axles in the road and had to be pulled out with heavy equipment. The road was unusable and had to be substantially re-built in order to resume the project. The road re-construction cost the project an additional $205,000.

Any future restoration projects at Tijuana Estuary will be faced with similar constraints. Initial phases of restoration identified in the Feasibility Study, should they be funded, will be confined to construction during non-breeding season.

The BA, which forms the basis of the FESA consultation, focuses on the potential effects of the project on threatened and endangered species. However, it also must provide an accurate description of existing conditions, direct and indirect impacts to sensitive habitats and T&E species, and mitigation for those impacts. A certified EIS, EA or Findings of No Significance (FONSI) are also required pursuant to NEPA.

CLEAN WATER ACT

The Clean Water Act (CWA) is the primary federal law governing water pollution. Its objective is to restore and maintain the chemical, physical, and biological integrity of the nation's waters by preventing point and non-point pollution sources, providing assistance to publicly-owned treatment works, and maintaining the integrity of wetlands. It is administered by the EPA and the USACE, in coordination with state governments, such as the RWQCB.

Technically, the act is named the Federal Water Pollution Control Act (FWPCA). The first FWPCA was enacted in 1948, but took on its modern form when completely rewritten in 1972 in an act entitled the Federal Water Pollution Control Act Amendments of 1972. Major changes have subsequently been introduced via legislation action including the Clean Water Act of 1977 and the Water Quality Act of 1987.

Section 404 of the CWA establishes a program to regulate the discharge of dredged or fill material into waters of the United States,

including wetlands. Activities in waters of the United States regulated under this program include fill for development, water resource projects (such as dams and levees), infrastructure development, including but not limited to, such as highways, airports and mining projects. Section 404 requires a permit before dredged or fill material may be discharged into waters of the U.S., unless the activity is exempt from Section 404 regulation, such as certain farming and forestry activities.

The basic premise of the program is that no discharge of dredged or fill material may be permitted if:

- a practicable alternative exists that is less damaging to the aquatic environment or
- the nation's waters would be significantly degraded.

To apply for a permit, it must first be shown that steps have been taken to avoid impacts to wetlands, streams and other aquatic resources; that potential impacts have been minimized; and that compensation will be provided for all remaining unavoidable impacts.

Proposed activities are regulated through a permit review process. An individual permit is required for potentially significant impacts. Individual permits are reviewed by the USACE, which evaluates applications under a public interest review, as well as the environmental criteria set forth in the CWA Section 404(b)(1) Guidelines, regulations promulgated by EPA. However, for most

discharges that will have only minimal adverse effects, a general permit may be suitable. General permits are issued on a nationwide, regional, or state basis for particular categories of activities. The general permit process eliminates individual review and allows certain activities to proceed with little or no delay, provided that the general or specific conditions for the general permit are met. For example, Nationwide Permit #27 Aquatic Habitat Restoration, Establishment and Enhancement Activities is applicable to restoration projects that do not have significant impacts to waters of the U.S., including wetlands. States also have a role in Section 404 decisions, through state program general permits, water quality certification, or program assumption. In California, the RWQCB assumes the role of the EPA by enforcement of Section 401of the CWA Water Quality Certification.

Nearly every tidal and riparian wetland restoration project conducted in Tijuana Estuary and river valley required a Section 404 permit. Exceptions include early work on restoration techniques conducted by Dr. Zedler and her students and the Napolitano Project, which qualified for Nationwide Permit # 27.

Much like the FESA, application for a Section 404 permit requires extensive information on the proposed project, submitted in what is called an "Environmental Assessment" (EA). A USACE EA is different from an EA prepared under NEPA in that it is more narrowly focused, dealing only with issues for which the USACE has jurisdiction.

Information required for an individual Section 404 permit and EA include:

- Description of existing conditions, including vegetation communities;
- A certified EIS, EA or FONSI pursuant to NEPA;
- Potential for the occurrence of T&E species and other wildlife;
- Delineation of USACE jurisdictional habitats based on accepted USACE methodology;
- Quantification of impacts to USACE jurisdictional habitats;
- Proposed mitigation.

While the EA may include information on sensitive habitats that are outside of the USACE jurisdiction, it must contain a description of those jurisdictional habitats. For tidal restoration projects, the jurisdictional limit is the high tide line. All areas above that elevation are non-jurisdictional.

The potential occurrence of threatened and endangered species triggers consultation with the USFWS pursuant to the Fish and Wildlife Coordination Act. The USACE initiates consultation under this process.

Impacts to jurisdictional habitats, including wetlands, are subject to mitigation at ratios of 1:1 or higher. The USACE may require project modifications to avoid or minimize such impacts.

Section 404 permits contain conditions that dictate project implementation, including but not limited to, the timing of construction, construction methods, beneficial reuse and disposal of excavated sediments and soil, and success criteria for proposed mitigation. One of the most important lessons regarding Section 404 permitting during implementation of restoration projects at Tijuana Estuary is the need to build in adaptive management. As presented previously, not long after construction of the Model Marsh, winter storms coupled with high tides deposited sediment in the channels and on the marsh plain. Researchers at PERL declared that their experiments using replicate tidal creek networks were jeopardized as they no longer functioned as replicates and requested that remedial dredging be conducted to restore the channels to their original configuration. The scope of work developed to remediate this deposition was not pursued due to the amount of effort needed to collect the information needed to modify the existing Section 404 permit for the project.

Hindsight once again suggests that provisions for remedial dredging of the site should have been included in the original permit. However, such an adaptive management condition is complicated, as impacts from remedial dredging would likely require mitigation. Once upland areas are excavated and connected to the tide, they become jurisdictional. Impacts to marine invertebrates, fish and birds that colonized the restoration site would likely trigger mitigation at ratios of 1:1 or higher. If a restoration project, such as the Model Marsh, has not been constructed as mitigation, it is possible that the remaining undisturbed portion of the site could serve as the required mitigation for remedial dredging. Details on such credits would have to be negotiated with the USACE. Access, staging areas and disposal of dredged materials would need to be anticipated in advance and could result in further impacts requiring further mitigation. Potential impacts to threatened and endangered species would require re-initiation of FESA consultation which could result in project restrictions, especially breeding season restrictions. Work would be confined to the rainy season where storm events could repeat the initial sediment deposition. Restorations undertaken as mitigation would likely not qualify for such a permit condition unless they were intentionally overbuilt (exceeding acreage required by the USACE), which is unlikely given the expense of tidal wetland restoration.

As climate change and predicted sea level rise are anticipated in the design of tidal wetland restoration projects, methods for building resiliency are being examined. Intentional placement of fill in jurisdictional areas that could facilitate transgression to wetlands in the future, which has been suggested in some projects, would under existing USACE policy be considered an impact and require mitigation. Many restoration plans in the region are considering stockpiling excavated soils rather than disposing of them so that there is a source of native soil to re-use to raise marsh elevations in the future. Unless USACE regulatory policy changes in the future, those actions would also be

considered impacts. Restoration projects are sometimes required to include high tide refugia for such species as the light-footed Ridgway's rail. However, these refugia, by their legal definition, must be higher than the high tide line and, therefore, do not receive credit from the USACE should they be constructed for mitigation.

CLEAN WATER ACT SECTION 401 WATER QUALITY CERTIFICATION

The California State Water Resources Control Board (CSWRCB) through the local branch of the Regional Water Quality Control Board (RWQCB), has the authority to approve, with or without conditions, or deny projects that potentially impact wetlands and/or other waters of the state. Guidance on such actions include the state's Porter-Cologne Act through Waste Discharge Requirements (WDRs) to protect waters of the state and the federal CWA under Section 401.

In addition to the state directives to protect wetlands, the Basin Plan also directs the CSWRCB staff to use EPA CWA 404(b)(1) guidelines to determine circumstances under which the filling of wetlands may be permitted and requires that attempts be made to avoid, minimize, and only lastly to mitigate for adverse impacts.

California's jurisdiction to regulate its water resources is much broader than that of the federal government and extends to isolated, intrastate, non-navigable waters as "waters of the United States" in the absence of federal regulation. Porter-Cologne extends to "waters of the state," which is broadly defined as "any surface water or groundwater, including saline waters, within the boundaries of the state."

The CSWRCB website lists the following information that must be submitted for application for 401 certification:

- Description of site location, project purpose, and avoidance/minimization efforts;
- A full, appropriately detailed, and technically accurate description, including the purpose and final goal of the entire activity;
- Copies of all completed or draft federal, state, and local permits or agreements related to the project;
- Documentation of coordination with other agencies;
- A copy of CEQA compliance documentation (if available) and any other required environmental documents (required before any approval action);
- U.S. Environmental Protection Agency's 404(b)(1) guidelines analysis;
- Mitigation & Monitoring Plan, if required
- Any other information requested by the Water Board staff;
- Fee: A $500 base price fee is required for fill and excavation, dredging, and shoreline discharges in addition to assessments based on the size of the projects (maximum fees of $40,000).

Most of the restoration projects in the Tijuana River Valley and estuary have required both a USACE Section 404 permit and RWQCB 401 certification. Most 401 certifications have been fairly straightforward and include information provided for other permits. One project in particular, the Fate and Transport Project, entailed more complicated negotiations with the RWQCB. As presented previously, that project entailed the placement of 45,000 yd³ of sediment captured in the Goat Canyon sediment basins within the surf zone at Border Field State Park. Such an action is in direct conflict with the intent of 401 certification which is to protect the waters of the state from discharges that might prove to be impactive. Although the project was designed as an experiment within a National Estuarine Research Reserve, the RWQCB imposed permit conditions that made conducting the experiment difficult. Turbidity from placement of fine grained sediment in the surf zone could not exceed 20% above ambient or the project had to be modified or halted, despite the intent of the project to create turbidity to study its effects. Qualitative turbidity monitoring was required by a qualified observer from a high vantage point and quantitative turbidity monitoring was measured from a vessel using a laser diffraction particle size analyzer in five to 25 meters of water at 10 different sampling sites within one kilometer of the placement site. Turbidity exceeded greater than 20% using both methods on several occasions and the project was halted until the turbidity dissipated. This affected the overall design of the experiment which was designed to mimic natural discharge of sediment by the Tijuana River during a

winter storm event. This example points out the need for greater flexibility in permitting experiments in areas designated as research reserves.

CALIFORNIA COASTAL ACT

The California Coastal Commission (CCC) was established by voter initiative in 1972 (Proposition 20) and later made permanent by the Legislature through adoption of the California Coastal Act of 1976.

The CCC, in partnership with coastal cities and counties, plans and regulates the use of land and water in the coastal zone. Development activities, which are broadly defined by the Coastal Act, include, but are not limited to, construction of roads and buildings, divisions of land, and activities that change the intensity of use of land or public access to coastal waters. Such activities generally require a Coastal Development Permit (CDP) from either the CCC or the local government pursuant to a certified Local Coastal Program (LCP).

Development within the coastal zone may not commence until a coastal development permit has been issued by either the CCC or a local government that has a CCC-certified LCP. After certification of an LCP, coastal development permit authority is delegated to the appropriate local government, but the CCC retains original permit jurisdiction over certain specified lands (such as tidelands and public trust lands). The CCC also has appellate authority over development approved by local governments in specified

geographic areas as well as certain other developments.

Along with the Bay Conservation and Development Commission (BCDC), the California Coastal State Conservancy, and the National Estuarine Research Reserve System (including the Tijuana River NERR), the CCC is one of California's four designated coastal management agencies for the purpose of implementing the federal Coastal Zone Management Act (CZMA), administered by the National Oceanic and Atmospheric Administration (NOAA), in California. Under California's federally-approved Coastal Management Program, the CCC manages development along the California coast except for San Francisco Bay, where the San Francisco Bay Conservation and Development Commission oversees development. The Coastal Conservancy leads efforts related to coastal restoration, while the NERR system is focused on research that can improve management and enhance education and outreach related to the state's estuaries.

One of the most significant provisions of the federal CZMA gives state coastal management agencies regulatory control (federal consistency review authority) over all federal activities and federally licensed, permitted or assisted activities, wherever they may occur if the activity affects coastal resources. Examples of such federal activities related to coastal wetland restoration projects include USACE CWA Section 404 permits; certain USFWS permits and roadway improvement projects.

Coastal wetland restoration projects that have been constructed within the boundaries of the USFWS Tijuana Slough National Wildlife Refuge (Tidal Linkage and Napolitano Restoration) required federal consistency determinations by the CCC. Restoration and research projects conducted within the boundaries of BSP (Model Marsh, Goat Canyon sediment basins and Fate and Transport projects) required CDPs from both the CCC and the City of San Diego as the projects included features within the jurisdictions of both agencies.

The CCC is generally supportive of restoration activities, but often adheres to a strict interpretation of the Coastal Act as it pertains to impacts to existing wetlands. For example, the CCC does not consider construction or improvement of transportation projects, including roads, within the coastal zone that may impact sensitive resources, such as wetlands, allowable under the act. Accordingly, the CCC did not permit the entirety of the proposed Goat Canyon Enhancement Project. That project proposed construction of the sediment detention basins and the raising of Monument Road from the sediment basins to Monument Mesa, portions of which had been buried from years of deposition.

The CCC objected on the grounds that there would be impacts to high salt marsh from fill placed to raise the road. The City of San Diego allowed that action within their jurisdiction. Monument Road was raised until the boundary between the two jurisdictions, where it sloped downward to meet existing grade within the CCC jurisdictional limits. The result again illustrates the unanticipated effects of good intentions leading to bad results. Once the sediment detention basins were operational, relatively sediment-free water exiting the basins found the lowest elevation as a flow path. This lowest elevation was the section of Monument Road within the CCC jurisdiction which became a de facto channel conveying sometimes highly contaminated water from Tijuana via Goat Canyon Creek. This water often ponds in low spots in the road resulting in closure of BFSP due to potential risks to human health. In addition, water conveyed across the border by Yogurt Canyon frequently ponds near Monument Mesa and poses similar risks. Denying the raising of the entire length of the road resulted in the unintended consequences of frequent closure of BFSP to the public. Employees of USFWS and CSP that must access this area of the park must wash.

NATIONAL HISTORIC PRESERVATION ACT OF 1966

Section 106 of the National Historic Preservation Act of 1966 (NHPA) requires federal agencies to take into account the effects of their undertakings on historic properties, and afford the Advisory Council on Historic Preservation (ACHP) a reasonable opportunity to comment. The Historic preservation review process mandated by Section 106 is outlined in regulations issued by ACHP.

The responsible federal agency first determines whether the proposed activity could affect historic properties. Historic properties are those that are included in the National Register of Historic Places or that meet the criteria for the National Register. If so, the federal agency must consult with the appropriate State Historic Preservation Officer or Tribal Historic Preservation Officer (SHPO or THPO) during the process. It should also involve the public, and identify other potential consulting parties. If it determines that it has no undertaking, or that its undertaking has no potential to affect historic properties, the agency has no further Section 106 obligations.

If the project could affect historic properties, the agency determines the scope of appropriate identification efforts and then identifies historic properties in the area of potential effects (APE). The agency reviews background information, consults with the SHPO or THPO and others, collects information from knowledgeable parties, and conducts additional studies as necessary. Districts, sites, buildings, structures, and objects listed in the National Register are considered; unlisted properties are evaluated against the National Park Service's published criteria, in consultation with the SHPO or THPO and any Indian tribe or Native Hawaiian organization that may attach religious or cultural importance to them.

If the agency finds that no historic properties are present or affected, it provides documentation to the SHPO or THPO and, barring any objection in 30 days, proceeds

with the project. If the agency finds that historic properties are present, it proceeds to assess possible adverse effects.

If the agency, SHPO or THPO agree that there will be no adverse effect, the agency proceeds with any agreed-upon conditions. If they find that there is an adverse effect, or if the parties cannot agree and ACHP determines within 15 days that there is an adverse effect, the agency begins consultation to seek ways to avoid, minimize, or mitigate the adverse effects.

The agency consults to resolve adverse effects with the SHPO or THPO and others, who may include Indian tribes and Native Hawaiian organizations, local governments, permit or license applicants, and members of the public. Consultation usually results in a Memorandum of Agreement (MOA), which outlines agreed-upon measures that the agency will take to avoid, minimize, or mitigate the adverse effects. In some cases, the consulting parties may agree that no such measures are possible, but that the adverse effects must be accepted in the public interest.

The presence of cultural resources, both historic and prehistoric, have affected restoration projects at Tijuana Estuary. Although the Goat Canyon Enhancement Project was a California State Parks project, the potential impacts to waters of the U.S., including wetlands, required a Section 404 CWA permit from the USACE. Thus, the USACE became the federal agency as defined under the NHPA. In addition, CSP

cultural resource staff took an active role in review of the project. Both agencies initiated consultation with SHPO. The project originally planned to transport sediment from the sediment basins to the beach via Monument Road. Although somewhat circuitous, this route would have allowed trucks to travel on a paved surface and avoid potential conflicts with equestrians on the horse trail. CSP records showed that the remains of some of the military installations that were located within BFSP near Monument Mesa were buried beneath Monument Road and it was feared that compaction from trucks filled with sediment could damage these buried resources. Thus, the project was forced to abandon this preferred route and use the horse trail road. As presented previously, this unpaved road could not support the weight of the trucks and sediment. The project had to be halted and the road substantially re-built at a cost of approximately $207,000 and impairment of the experiment.

Similarly, the Model Marsh project required a Section 404 permit and the USACE initiated consultation with SHPO. A condition of the 404 permit was that an archaeological monitor be present during all ground disturbing activities. In December 1999, just prior to the completion of project grading, the monitor discovered a prehistoric shell midden near the eastern end of the constructed marsh. Work was halted in this area while a data recovery plan was completed, accepted by SHPO and implemented. The project was completed one day before the USFWS-imposed end date of February 15, 2000. The data recovery plan and its implementation cost the project an additional $126,000. During the time it took to develop and implement the recovery plan, the marsh plain dried and became hypersaline. Even after tidal influence was introduced, the soils remained hypersaline. Approximately 5,000 salt marsh seedlings grown by PERL as part of the experiments incorporated into the project died from the hypersaline conditions and replacement plants had to be propagated resulting in increased cost and delays in the experiment.

In summary, all restoration practitioners recognize the importance of environmental regulations and their role in protecting and preserving the valuable aquatic resources of California and the country. Unfortunately, not all consequences of permit conditions and their implementation can ever be foreseen, and resource agencies are typically understaffed, so the process of acquiring permits and later modifying those permits can be tedious. One of the main lessons learned from restorations at Tijuana Estuary, as well as those elsewhere, is that a project never behaves exactly as planned. Almost immediately after a restoration is built and is connected to tidal influence, it starts to deviate from the plan. In some cases that deviation might result in an unexpected improvement to the project; in others a detriment, but it almost certainly won't behave exactly as planned. That's where adaptive management, or adaptive restoration should be applied to allow mid-course corrections. To the extent possible, certain processes common to all tidal restoration projects, such as scour and deposition of sediment, should be anticipated and flexibility incorporated into permits. While this may result in complex trade-offs, the ability to correct the course of the intended design can prove to be very valuable in the long term.

CHRIS NORDBY is a wetland ecologist with over 40 years of experience in designing, constructing and monitoring coastal wetland restoration projects. He has a BA in Zoology from the University of Northern Colorado and a MS in Ecology from San Diego State University. Mr. Nordby conducted much of his research in coastal wetlands at Tijuana Estuary in southwestern San Diego County where he served as manager of the Pacific Estuarine Research Laboratory (PERL) from 1985 to 1990. He has participated in numerous restoration projects in the San Diego area, including those at Tijuana Estuary, south San Diego Bay, Mission Bay, Los Peñasquitos Lagoon, San Dieguito Lagoon, San Elijo Lagoon and Aqua Hedionda Lagoon, as well as Ballona Wetland in Los Angeles County.

LITERATURE CITED

Boland, I. 2018. The Kuroshio Shot Hole Borer in the Tijuana River Valley in 2017–18 (Year 3): Infestation Rates, Forest Recovery and a New Model. Final Report for the U.S. Navy and U.S. Fish and Wildlife Service and Southwest Wetlands Interpretive Association.

California Regional Water Quality Control Board. 1994. Water Quality Control Plan for the San Diego Region (9) (with amendments effective on or before April 4, 2011).

Mahall, B.E and R.B. Parks. 1976. The ecotone between *Spartina foliosa* Trin. And *Salicornia virginica* L. in salt marshes of northern San Francisco Bay. I. Biomass and production. J. Ecol. 64: 421–433.

Mitsch W. J. and J. G. Gosselink. 1986. Wetlands. Van Nostrand Reinhold Company, Inc. New York. 539 pp.

Nordby, C. S. and J.B. Zedler. 1991. Responses of Fish and Macrobenthic Invertebrate Assemblages to Hydrologic Disturbances in Tijuana Estuary and Los Peñasquitos Lagoon, California. *Estuaries*, 14(1): pp. 80–93.

Nordby Biological Consulting and the Tijuana River National Estuarine Research Reserve. May 2014. South San Diego Bay Coastal Wetland Restoration and Enhancement Project. Year 2 Postconstruction Report.

Nordby Biological Consulting and the Tijuana River National Estuarine Research Reserve. July 2015. South San Diego Bay Coastal Wetland Restoration and Enhancement Project. Year 3(2014) Postconstruction Report.

Phillip Williams and Associates. 2000. Preliminary Draft, Goat Canyon Sediment Retention Basin Alternative Development.

Safrans, S.M., Baumgarten, S.A., Beller, E.E., Crooks, J.A., Grossinger, R. M., Lorda, J., Lingcore, T.R., Bram, D., Dark, S.J., Stein, E.D. and MacIntosh T.L. 2017. Tijuana River Valley Historical Ecology Investigation. Prepared for the California State Coastal Conservancy.

Seamans, P. 1988. Wastewater Creates a Border Problem. Journal of the Water Pollution Control Federation. 60:1798–1804.

Tierra Environmental Services. 1999. *Cultural Resources Evaluation of Site CA-SDI-14,831 for the Tijuana Estuary Restoration Project/Model Marsh, Border Field State Park, California.* Prepared for the California State Coastal Conservancy.

Tierra Environmental Services. 2008. *Tijuana Estuary–Friendship Marsh restoration Feasibility and Design Study.* Prepared for the California State Coastal Conservancy.

Tierra Environmental Services. 2002. Biological Assessment for the Goat Canyon Enhancement Project.

U.S. Army Corps of Engineers and U.S. Environmental Protection Agency. 1991. Evaluation of Dredged Material Proposed for Ocean Disposal—Testing Manual.

U.S. Army Corps of Engineers and U.S. Environmental Protection Agency. 1998. Evaluation of Dredged Material Proposed for Discharge in Waters of the U.S.—Testing Manual. Inland Testing Manual.

Warrick, J.A. 2013. Disposal of Fine Sediment in Nearshore Coastal waters. Journal of Coastal Research. Vol. 29. Issue 3. pp 579–596.

Zedler, J.B., C. Nordby, and P. Williams. 1979. Clapper rail habitat: requirements and improvement. Project final report, U.S. Department of the Interior, U.S. Fish and Wildlife Service, Sacramento Endangered Species Office, Sacramento, California.

Zedler, J. B., J. Covin, C. Nordby, P. Williams and J. Boland. 1986. Catastrophic Events Reveal the Dynamic Nature of Salt Marsh Vegetation in Southern California. *Estuaries* Vol. 9. No.1, pp75–89.

Zedler, J.B. and C.S. Nordby. 1986. *The Ecology of Tijuana Estuary, California: An Estuarine Profile*. NOAA Office of Coastal Resource Management, Sanctuaries and Reserves Division, Washington D.C., 104 pp.

Zedler, J.B., C.S. Nordby, and B. Kus. 1992. *The Ecology of Tijuana Estuary: A National Estuarine Research Reserve*. Pacific Estuarine Research Laboratory (PERL). San Diego State University, San Diego, CA.

Zedler, J.B. ed. 2001. *Handbook for Restoring Tidal Wetlands*. CRC Marine Science Series, CRC Press, Boca Raton, FL.

Acknowledgements

The following people provided input and guidance that greatly improved this document:

John Boland
Boland Ecological Services

Brian Collins
U.S. Fish and Wildlife Service, San Diego Bay and Tijuana Slough National Wildlife Refuges

Jeff Crooks
Tijuana River National Estuarine Research Reserve (TRNERR)

Greg Gauthier
California State Coastal Conservancy

Sam Jenniches
California State Coastal Conservancy

Chris Peregrin
California State Parks, TRNERR

Victoria Touchstone
U.S. Fish and Wildlife Service, San Diego National Wildlife Refuge Complex

Mayda Winter
Southwest Wetlands Interpretive Association

Joy Zedler
University of Wisconsin–Madison, Emerita

—

Monica Almeida
TRNERR, provided graphics support

Jan Carpenter Tucker
Night Star Publisher, designed and published this book